LIFE AS NO ONE KNOWS IT

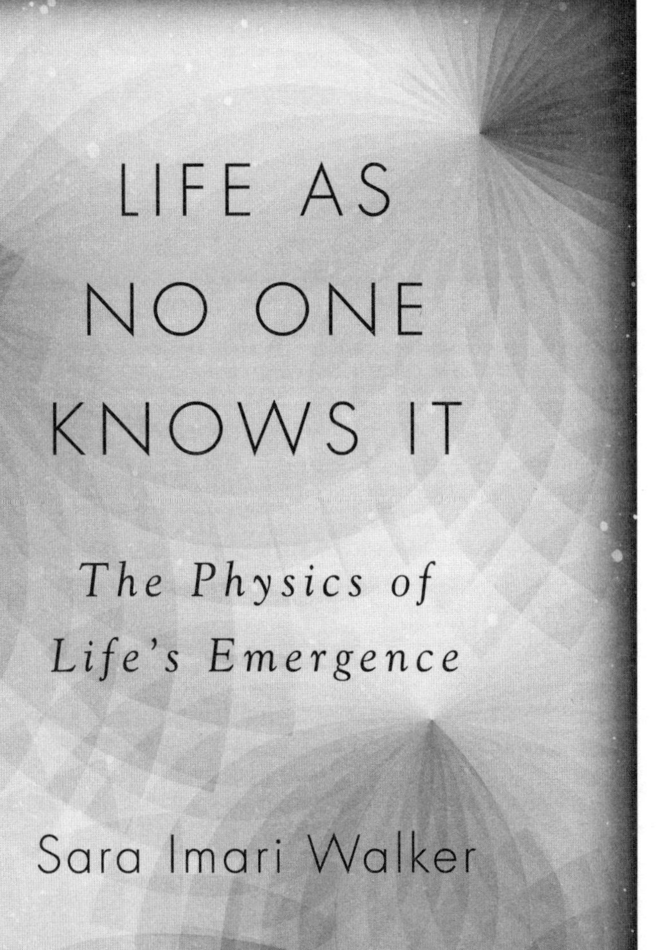

LIFE AS NO ONE KNOWS IT

The Physics of Life's Emergence

Sara Imari Walker

The Bridge Street Press

THE BRIDGE STREET PRESS

First published in the United States of America in 2024 by
Riverhead Books, an imprint of Penguin Random House LLC
First published in Great Britain in 2024 by The Bridge Street Press

1 3 5 7 9 10 8 6 4 2

A CIP catalogue record for this book
is available from the British Library.

Hardback ISBN: 978-0-34912-824-5
Trade paperback ISBN: 978-0-34912-823-8

Some of the material in Chapter 6 "Planetary Futures" previously
appeared in different form in *Noēma* (April 2023).

Book design by Meighan Cavanaugh
Printed and bound in Great Britain by Clays Ltd, Elcograf S.p.A.

Papers used by The Bridge Street Press are from well-managed forests
and other responsible sources.

The Bridge Street Press
An imprint of
Little, Brown Book Group
Carmelite House
50 Victoria Embankment
London EC4Y 0DZ

An Hachette UK Company
www.hachette.co.uk

www.littlebrown.co.uk

To my lineage . . .

with the hope we may one day come

to understand what we are

CONTENTS

LIFE AS NO ONE KNOWS IT

One

WHAT IS LIFE?

Have you ever wondered what makes you alive? What makes anything alive?

At the 2012 meeting of the American Chemical Society, in a session on the origin of life, Andrew Ellington proposed a radical theory: "Life does not exist." Andy is a chemistry professor from the University of Texas at Austin, and this was the first slide of his presentation on RNA* chemistry and the origin of life. His idea left me incredibly perplexed.

I was perplexed because I probably should have agreed with Andy. But I don't. When I attended Andy's lecture I was pretty sure I was alive, as I am now. You're probably confident you are

---

*RNA is an acronym for ribonucleic acid, an important biomolecule believed by some to play a critical role in the origin of life due to its function as an information-bearing molecule and a chemical catalyst.

alive too. Haven't you spent your whole life, well, living? Being alive matters. It's very different from not being alive.

Yet despite our natural confidence in our own existence, some scientists challenge it and argue that life may be just an illusion or epiphenomenon, explainable by known physics and chemistry. Physicist and public intellectual Sean Carroll is one such individual. In a crowded evening lecture on the Arizona State University campus where I work, I was aghast in my seat as Sean stated how the equations of particle physics are sufficient to explain the existence of all matter—including you and me. Jack Szostak, a Noble Prize winner, holds a similar view, arguing that the focus on defining life is holding us back from understanding life's origin. According to Jack, the closer you look at any of the "defining" properties of life, the more the boundary between life and nonlife blurs.[1]

As a child, I remember trying to take an insect apart and then failing to restore it to its original state. I was too surprised at the time to even feel upset. We are all familiar with how life cannot be reduced to its parts, be they elementary particles, atoms, or even molecules. Perhaps it is easiest to take the view, as Andy, Jack, and Sean do, that life is not a property of its parts, and that therefore we don't need to worry about defining it. If true, it follows that all we need to understand what life does and how it emerges is to understand those parts.

In my training as a theoretical physicist, I was taught to believe that life was not a conceptually deep scientific problem. Instead, the most fundamental concepts regarding the nature of reality were what other physicists had studied—things like space, time, light, energy, and matter. Indeed, the successes of physics have been nothing short of profound: over the short span of the last four hundred years,[2] we have gained a deep understanding of how our universe works. We have even defined what we mean by "universe." At the very small scales, we understand much about the elementary constituents of all matter. At the very largest scales, we can take photos of distant galaxies whose light took more than 13.5 billion years to reach our telescopes.

Yet the origin of life remains one of the greatest puzzles in science. Physics, as we collectively understand it at this moment in history, provides a fundamental description of a universe devoid of life. It's not the universe I live in, and I bet you don't live there either.

But if life does exist, what is it?

What arc wc?

If Vitalism Is Dead, Maybe You Are Too

In stark contrast to the views of modern physicists and chemists, scientists used to believe life exists as a separate category from matter.

Animated matter was believed to be imbued with a "vital" force,[3] sometimes referred to as the élan vital. Aristotle called it entelechy; Gottfried Wilhelm Leibniz called it monads.[4] They were both describing, as many have, a unique quality found only in living entities that directs living behaviors, such as the development of an embryo, regeneration of a lost limb, or any of the other purposeful activities that seem uniquely characteristic of life. This concept of being alive is somewhat akin to the religious concept of a soul, and some have even called it that. Whatever you call it, we think these features are unique to life because we do not observe them in nonliving things. A rock does not restore its original shape when cut in two, but a planarian worm can and does. Vitalism, as the scientific movement came to be called, was driven by the idea that what makes matter come alive cannot be described mechanically and is therefore not material.

While modern materialists like Andy, Sean, and Jack regard the known properties of matter as sufficient for explaining life, the vitalists had quite the opposite view. They believed that life does, in fact, exist, but it *cannot* be explained in terms of the properties of matter.

Often, the idea of a vital principle was discussed in terms of a life energy or vital spark that could animate even dead matter. If this sounds a bit Frankensteinian, that's because it is. Mary Shelley was just twenty-one when her famous novel *Frankenstein* was published in 1818.[5] When she penned the book, she was reflecting on the leading science of her day, particularly theories on the soul, what makes us alive, and how we might reanimate the dead with electricity.

Mary was influenced by the contemporary work of Luigi Galvani, work later carried on by his nephew Giovanni Aldini.[6] These two attempted to animate body parts through electrostimulation; they would stimulate dead frogs' legs with electric shocks to make them "dance." Mary was also reportedly inspired by Erasmus Darwin, grandfather of the more famous Darwin, Charles. The elder Darwin wrote on the topic of spontaneous generation, citing how inanimate materials could spontaneously become animated in water warmed by sunlight.[7]

In Mary's novel, the body parts of the recently deceased could be reanimated by electrocution if they were properly wired together. This is how her lead character, Dr. Victor Frankenstein, made his "living" monster from dead bodies. If we were to try to write down laws of physics that could explain Dr. Frankenstein's unique insight in animating the monster, we might conjecture he had discovered a life "force" that dissipated slowly after death, and that whatever substance this force was made of was strongly coupled to electromagnetism. It's a bit of an odd set of properties Dr. Frankenstein discovered in his fictional universe, but matter in our real universe has many strange properties too. Our universe is weird when you start to understand it (in fact, it gets odder the more you think you understand it).

We might imagine a consistent physics that Dr. Frankenstein tapped into that explains life. It just so happens that whatever physics he was onto is not the physics that describes our real universe. The real physics underlying life might be even stranger still. While right now we do not understand what principles govern life, they may one day be as obvious to subsequent generations as the curvature of space-time or the existence of particles of light (photons) are to us now.

Many of the vitalists thought life could not be produced by things that were not already themselves alive. Living matter was special because its parts were special, carrying some of that requisite élan vital. Thus, living things were necessary to make more living things; even Dr. Frankenstein had to make his monster from once living parts that were only recently deceased.

Around the time of the publication of *Frankenstein*, the idea that life is necessary to produce the stuff of life was already beginning to lose popular support within the scientific community. In 1828, Friedrich Wöhler synthesized urea, an organic molecule found in urine, from two other simple molecules, cyanic acid and ammonium. Friedrich's experiment showed that biologically derived molecules do not carry any sort of life force. The component parts of living matter are no different from those of nonliving matter. In experiments like these, scientists have repeatedly demonstrated that there is nothing that separates the properties of nonliving chemistry from living chemistry: the former can easily be transformed to the latter under appropriate conditions. The sharp boundary between nonlife and life started to blur as humanity began to understand more about chemistry and the physics underlying it. Sean, Jack, Andy, and many others who hold similar views are right . . . to a degree.

In fact, as much as we have looked, we have found the transformation from a nonliving to a living substance is not excluded by any known law of physics or chemistry. There is no life conservation law that says life cannot be created or destroyed. Of course, this is obvious, because we know organisms are born and die—but sometimes it is the most obvious observations that are the hardest to explain scientifically, and harder still to turn into mathematical law.

What modern science has taught us is that life is not a property of matter.

Physicists and chemists see very intimately what the rest of us who think life exists cannot: there is no magic transition point where a molecule or collection of molecules is suddenly "living."

Life is the vaporware of chemistry: a property so obvious in our day-to-day experience—that we are living—is nonexistent when you look at our parts.

If life is not a property of matter, and material things are what exist, then life does not exist. This is probably the logic Andy was going for.

Yet here we are.

The Paradox of Defining Life

Look around you right now. I'll bet you can confidently catalog what is living and what is not. I could quiz a four-year-old, and they would probably corroborate your classification. In fact, small children seem rather good at telling the difference between living things and inanimate objects with little explicit direction. Once a child is taught that plants are alive, they can readily tell those apart from nonliving things and act accordingly—for example, not stomping on the flower bed but perhaps jumping in the mud instead. Kids can also misclassify life by assuming inanimate objects, like a favorite stuffed animal, are life. Are they wrong in doing this, or are we wrong in correcting them?

Whether we are teaching the correct story to our children assumes that we have it right ourselves. It seems so obvious to us that some things are alive and others are not that "I'll know it when I see it" has become a popular mantra at professional astrobiology conferences. I could count more times than I have fingers and toes how often I've heard this said in lectures by distinguished scientists. Typically, it is meant in jest about the fact that defining life is hard. Naturally this is almost invariably followed by the speaker attempting to define life.

The idiom "I'll know it when I see it" takes its popular use from the 1964 United States Supreme Court case *Jacobellis v. Ohio* in a statement made by Supreme Court justice Potter Stewart regarding the subjectivity of deciding whether something constitutes pornography.[8] The point made in the court case was that there is no objective definition for pornography—whether something is considered pornography or art is determined by the opinion of the viewer. Just as with defining pornography, current definitions of life are based on our subjective understanding. There are no objective criteria that are universally accepted by scientists, or anyone else for that matter.

This is not for lack of effort. There are dozens of definitions proposed for life. While it might seem like life should be easy to define, it is not.[9] For any definition there are always exceptions.

Consider, for example, a definition that includes "the ability to eat, metabolize, excrete, breathe, move, grow, reproduce, and be responsive to external stimuli." You surely are doing at least a few of these things right now. If you opened an introductory biology textbook, you would likely encounter a similar list without having to read past the first chapter. Biologists have gotten quite good at describing life on Earth. However, this description doesn't necessarily help us understand life as a general phenomenon in our universe, or to understand the artificial life and intelligences we are creating. Using this defining list, any aliens visiting the Earth might assume cars are the dominant life-form. Despite the way many of us like to name our cars or assign them a gender (my children named our car Little Blue), we usually view this as anthropomorphizing and typically do not consider our cars to be examples of life.

The potential for misclassification of cars as life was pointed out by Carl Sagan, the acclaimed science communicator and an early pioneer in astrobiology, in a colorful essay he penned about the vexed problem of defining life. As Carl pointed out in reference to the textbook definition, "Many such properties are either present in machines that nobody is willing to call alive, or absent from organisms that everybody is willing to call alive."[10] I disagree with Carl's assertion that we must exclude machines, but I do agree with his other key philosophical observation: one of the great paradoxes in attempts to define life—and a key reason why our definitions continue to fail us—is that even if we start from reasonable assumptions, we do not get what we feel are intuitively reasonable answers. We end up with cases we want excluded being included, and cases we want included being excluded.

One of the most popular definitions for life circulating in scientific communities is that "life is a self-sustaining chemical system capable of Darwinian evolution." This definition was first developed in an exobiology* discipline working group organized by John Rummel, who at the time was manager of NASA's Exobiology Program. Although the definition was developed under

*"Exobiology" is an alternative term for the branch of research also known as astrobiology. While both terms were used in the early days of the field, as the science has matured and become a more serious research endeavor with federally supported research programs, "astrobiology" became more commonly used, but both are still in use and can be used interchangeably.

the auspices of a NASA working group, it is by no means the official NASA definition, as is sometimes claimed in popular news outlets. It's not even necessarily a widely accepted definition.

Gerald Joyce, a highly respected RNA chemist, has been misaccredited as the sole author of this definition, which he himself will point out. This is because in 1994 he wrote about the working group's definition in the foreword of a book titled *Origins of Life: The Central Concepts*.[11] RNA, short for ribonucleic acid, is believed to have played an early role in the origin of life, at least by those scientists who adhere to the RNA World hypothesis.[12] This model proposes that life started with RNA, as opposed to DNA (deoxyribonucleic acid), as the genetic material. During the last few decades, the RNA World has been considered the leading hypothesis for the origin of life on Earth. Gerry studies the in vitro evolution of RNA molecules to understand potential mechanisms of chemical evolution in the putative RNA World. He and his colleagues have demonstrated amazing capabilities in RNA, such as the ability for self-replication and evolution under controlled conditions. Because RNA chemistry isolated and evolved in a test tube can fulfill so many functions of living biochemistry, it is understandable one might arrive at a definition of life as a self-sustaining chemical system capable of Darwinian evolution.

A lot of people use this definition and like it. But at the same time, like all definitions for life, it fails to deliver on solving the hardest problems. It has not allowed us to prove a mechanism for the origin of life nor design new instruments to search for and measure the presence of alien life on another world.

Let's first focus on the notion of "self-sustaining." In vitro evolution proceeds by taking a sample from one test tube and transferring it to another, and then another, to simulate successive rounds of evolutionary selection. Because evolution occurs only by intervention of the experimenter, it requires continual human input to sustain an evolving population of RNA.* Most people do not feel comfortable referring to these systems as "life" because they are not self-sustaining and they require human intervention. Most people thus take the stance, like the NASA working group did, that such systems should be excluded.

However, this reasoning also excludes parasites: instead of depending on an experimenter and a test tube, parasites depend on their host organism to survive and reproduce. A fascinating example is the parasite *Ophiocordyceps unilateralis*, more popularly known as zombie ant fungus, a species of fungi that infiltrates and controls the brains of its host ant, leading the ant to abandon its colony as part of the life cycle of the fungus. These ants, piloted by fungi, are referred to as zombies because they are literally that: while animated, they are not in control. Many of us would consider parasites like *Ophiocordyceps unilateralis* alive, but it is not self-sustaining without the ant. And indeed, is the ant even really alive when it is being piloted? It requires the fungus at that point to live and to behave.

---

*In vitro evolution works by allowing RNA molecules to replicate in a test tube. Those that are the best replicators, i.e., those that are most "fit" in the terminology of evolutionary theory, will take a larger share of the resources and therefore be more abundant. A small sample of material is taken from the test tube and transferred to another with fresh resources, and exponential growth is allowed to happen again. Over successive iterations, the "fittest" molecules can be amplified in the population—that is, they are selected above the rest. In this way, evolution is not an autonomous process but requires an agent (the experimenter) to drive it.

Then there's the vexed issue of what defines a "self." Our ability to sustain our human selves is deeply dependent on others. Try to imagine how long a typical human, accustomed to life as enabled by technology and modern society, could survive in the wild on their own. For most of us this might be only days or weeks. This is not unlike the cells within your body, which would be hard pressed to survive outside the context of you as a living organism. Obviously, there are notable exceptions, like human alpinists who have trained to survive under some of the most harrowing conditions imaginable, or an untrained individual who happens to pick up the skills quickly enough to survive and has just the right amount of luck. But such exceptions are not the rule in modern societies. Are we less alive now as individuals than we were in the past, because once upon a time we were more self-sustaining as individuals (or at least lived in smaller self-sustaining groups)? Are only societies alive now, because they are the things that are self-sustaining?

Next, there's the concept of "Darwinian evolution," which many would argue is *the* defining feature of life. In general, Darwinian evolution applies only to things that can reproduce. Reproduction allows traits to be inherited by offspring from their parents, and selection on these inherited traits can cause populations to change over time. However, by this definition an astronaut in space is not alive and neither is your grandmother, because neither is capable of reproduction. Some organisms, like worker bees or mules, never have the chance to reproduce because they are literally unable to. Are these individuals *never* alive?

You meet all kinds of people at astrobiology conferences, and some love to challenge your ideas and have you challenge theirs.

Lucas Mix, an ordained priest and astrobiologist, is among this crowd. He is open-minded enough to really engage with things he disagrees with. Lucas makes a very clear point in what he calls Darwin life.[13] He argues that definitionally it applies only to populations, because populations are the only things that can evolve. I am not a Darwinian system, and neither are you (I mean the human part, not the large populations of microbial cells within our bodies, which can and do evolve). This means that you as an individual are not life and neither was Albert Einstein, but together you might constitute an example of life as members of the same evolving population. If we read too literally into the NASA definition, no individual is life.

In any definition, some things we meant to include will always fall outside the boundary of what counts as life. However, a real understanding of what life is should not depend on how comfortable we are with what will ultimately be considered "alive" or "life." Neither of these are precisely defined scientific terms. We don't yet have a general understanding of the category of things that we should group together and call "life." Therefore either our categorization is wrong or life is not something to be categorized.

Why Life?

If you visit any physics department and hang around the students long enough, you will invariably join a discussion about how all the "great" physicists became great by their mid-twenties. Newton was just twenty-three years old when he discovered the universal law of gravitation. Einstein was twenty-six when he had his "annus mirabilis," producing groundbreaking work on the photoelectric effect, Brownian motion, special relativity, and the equivalence of mass and energy. Given the youth of many of the individuals among history's greatest physicists, it is perhaps not surprising that some might think all the great discoveries in physics should also happen while societies themselves are still young, and there might even now be nothing very deep left to discover.

As you can imagine, this puts a bit of pressure at a young age on those among us who want to make significant contributions to the field of theoretical physics. It makes many physicists conservative in the problems we approach early in our careers. I, too, thought fundamental problems were what physicists said they were, so I was not very interested in the origin of life when my PhD adviser, Marcelo Gleiser, a theoretical physicist and cosmologist, suggested that I work on it. In fact, I spent the first three years of my PhD fairly resistant to the idea! I thought I would work on the origin of life for a time, but then get back to the business of being a real physicist and studying real physicist things. However, as my time in graduate school progressed, I became more and more intrigued by the origin-of-life problem. It had none of the baggage or dogmas of conventional theoretical physics and held all the things that attracted me to studying physics in the first place: in origin-of-life research, the ideas were on such tenuous theoretical footing that one could build a new understanding of reality from the bottom up, just as Isaac Newton had done for motion and gravity, James Clerk Maxwell for light, Charles Darwin for evolution, Emmy Noether for symmetries and conservation laws, and Werner Heisenberg and Erwin Schrödinger for elementary particles. As a PhD student, this was a risky thing to decide to spend your career on.

A challenge (as always) was: Where to start in uncharted territory? I was pointed repeatedly to the work of the now controversial, famous physicist Erwin Schrödinger, who played a foundational role in the development of quantum mechanics. In 1943, Erwin, then director of Theoretical Physics at Trinity College in Dublin, Ireland, delivered a series of lectures at the Dublin Institute for Advanced Studies on the topic of life, which was published in the subsequent year as a highly influential book. He aptly titled his lectures "What Is Life?"[14] Reading the book of lectures was refreshing for me as a student, not necessarily because of its content, but because of its validation of life as a problem for the physicist's conception of nature. Erwin argued that life should obey mathematical laws that are simple to describe and that are mathematically elegant.

Importantly for our discussion in this book, what is significant is not that ideas can be written in math. Instead, it is that the abstraction we uncover—which may be written in the form of mathematical laws that explain life—should have high explanatory power. In the philosophy of science, there is a lot of debate about what constitutes "explanation," but in general we might consider the explanatory power to be higher for theories that clarify more facts about the world, change surprising facts into everyday knowledge, make better (more accurate) predictions, have fewer assumptions, are testable by more observations or experiments, and as the famous Occam's razor holds, they should also be based on a relatively simple set of elements. In addition to these criteria, the quantum physicist David Deutsch adds that good theories are "hard to vary."[15] That is, a key feature of a good theory is that if you change the details even slightly it is no longer explanatory.

My view is that physics is not a discipline about particles and fields, galaxies and black holes or condensed matter—although those may be things some physicists study. Instead, physics is a way of approaching the world by developing abstract descriptions of nature that possess, in their abstraction away from the everyday objects of our experience, the highest explanatory power. In this book we are after a theory with high explanatory power that will provide new, deeper insights into the nature of life. Incidentally, the fact that explanatory theories work in physics, and science more broadly, is itself an important clue into the nature of life. Life is, after all, the only kind of physical system we know of that can write down explanations like the laws of physics.

Although just about every physicist's dream is to come up with new mathematical theories for how reality works, Erwin avoided mathematical explanation to make his "What Is Life?" lectures more accessible. They attracted an audience of several hundred, and Erwin wrote in the book version that the audience was warned that "the subject-matter was a difficult one and that the lectures could not be termed popular, even though the physicist's most dreaded weapon, mathematical deduction, would hardly be utilized." My guess is he knew enough to know that had he tried to explain life by mathematical deduction at that point in time, he would have failed. Maybe he did not want to admit this.

Erwin framed his question about life, assuming a well-defined boundary is significant. He asked, "How can the events in space and time which take place within the spatial boundary of a living organism be accounted for by physics and chemistry?" Here was his first major misstep, dooming him and many generations of physicists to not find the answer. If there is a fundamental unit of life, it is not an individual organism, because no individual is isolated from the evolutionary chain of events that produced it.

Erwin introduced a new concept to partially explain life, which he called negative entropy (later shortened to "negentropy" by another well-known physicist, Léon Brillouin, in his own writing on related topics).[16] Erwin drew inspiration from the second law of thermodynamics, which states that the entropy of an isolated system should always increase. The three laws of thermodynamics were developed in the 1800s by Sadi Carnot, Ludwig Boltzmann, and others, who were able to codify the relationship between heat and energy. In these laws, entropy is a way of measuring how disorderly things are. You can also think of it as a measure of the unavailability of energy for doing useful things. According to physicists, maximum "disorder"—aka maximum entropy—means a system can do nothing more of interest and therefore it is deadly boring.

For living things to, well, live, they need to harness energy from their environment—that is, they need to eat and then be able to metabolize what they have eaten to turn it into energy and biomass. The banana bread I had for a snack this afternoon is currently being metabolized in my stomach as I type this. This in effect decreases the entropy of my environment, creating more order instead of disorder. Because acquiring energy in this way is so central to what life must do to persist, many have confused this with what life is.

Erwin asked, if entropy always increases, then how do we exist? This is the crux of what is sometimes called Schrödinger's paradox: the fact that living organisms persist even though the second law suggests we should not.

As far as paradoxes go, this one is easily resolved. One need only recognize that the second law is a statistical and not an exact statement. Instead of thinking about a box with a gas of particles inside it, physicists will think about many identical boxes with the same kinds of particles inside them. There is nothing to prohibit a single box from violating the statistical trend of the rest, at least for a short time. If you had ten boxes with particles zigging and zagging about like billiard balls, one box could—by random chance—arrive at having all the particles in the same corner. This, for physicists, would be considered an ordered, low entropy state because there are fewer ways to describe the box with all the particles in the same corner than there are to describe the multitude of ways the box could be with particles zigzagging in random directions. Perhaps our biosphere is an example of a system in a state of negative entropy, only temporarily averting the second law by fluctuating into an unlikely ordered state. Indeed, this would be observationally consistent with what we know of the distribution of life in the universe—we appear to be the only planet that has undergone such a large, rare fluctuation to order and to life, as far as we know right now. A further reason there is no paradox here is that life on Earth is part of a larger open system, therefore statistically entropy can continue to increase overall in the universe despite the temporary fluctuation that is life as we know it.

In inventing the idea of negative entropy, Erwin did not explain life. He explained only how life could be compatible with the known laws of physics.

Local systems that maintain low entropy by harvesting external free energy (as we do from the Sun) and ejecting waste heat (as our biosphere does, in the form of radiation) are known as dissipative structures, described by Ilya Prigogine in work that won him the Nobel Prize in Chemistry in 1977. Ilya discovered how chemical systems that exchange energy and/or matter with their environment exhibit the emergence of different kinds of organization. Examples range from turbulent flows to convection cells to the patterned spots on a leopard. Ilya was a larger-than-life personality and built up an almost cultlike following based on his conviction that dissipative structures offer a general theory to explain much of the natural world, including life.

Yet there are plenty of examples of dissipative structures that most would not consider "living." Take for example the Great Red Spot of Jupiter. This storm is driven by the disequilibria conditions created when warm, moist air meets cold, dry air, and it has persisted as an organized structure in the atmosphere of Jupiter for nearly two hundred years.[17] There are somewhat similar dissipative structures on the surface of the Earth too—what we call hurricanes, which like the Great Red Spot are storms with high winds that self-organize into a temporarily persistent spiral shape. You would probably be just as hesitant to call a hurricane an example of life on Earth as you would be to call the Great Red Spot on Jupiter an example of alien life.

Unique to Earth (as far as we currently know) are dissipative structures also visible from space that are indeed examples of life: these we call cities. Cities provide an interconnected web of lights visible beyond Earth because of the electromagnetic radiation they give off at night. Like storms, they "self"-organize, only under far-from-equilibrium conditions. Hurricanes and cities are formed by some of the same physics: they are both examples of states of organization maintained far from equilibrium. Many things we observe can be explained in terms of the exchange of energy and work to construct organized assemblies of matter.

Dissipative structures include examples where life exploits the fact that open, far-from-equilibrium systems can locally maintain order. It is necessary that this be true for life as we know it to exist, but it does not explain *why life exists*. It does not explain the key way in which storms and cities differ. The current Great Red Spot has no memory of the multitude of similar systems that may have preceded it over the several-billion-year history of Jupiter; studying the storm won't allow us to extract a detailed history of storms that persisted in Jupiter's past. By contrast, cities are the direct consequence of evolutionary processes that began on Earth more than 3.7 billion years ago, with the gradual acquisition of information leading to the emergence of what we call cellular life, multicellular life, societies with language, and eventually the artifacts of those societies that we call cities. This memory is encoded in the very existence of a city, and you can find evidence of this history by peeling back the different architectural and biological layers much like a palimpsest. Cities do not spontaneously fluctuate into existence because a disequilibrium exists; they require a long causal chain of events—a lineage, as I'll describe throughout this book—for the universe to construct them. The memory that leads to the construction of a city is acquired over eons via selection;

it is not spontaneously self-organized in space, but instead assembled across time.

As hard as we look, there is no evidence that life violates any of the known laws of physics. But being consistent with known laws of physics does not mean life is explained by those laws. In *What Is Life?*, Erwin wrote:

> Living matter, while not eluding the laws of physics as established up to date, is likely to involve other laws of physics hitherto unknown.

That was over seventy-five years ago. To this day, despite the efforts of generations of talented scientists, we cannot derive life from the known laws of physics, even if we are pretty sure it must be consistent with them.

From Atoms to Agents

You may be wondering what any of the foregoing discussion has to do with the actual experience of being alive. Perhaps it sounds like scientists are focusing on all the wrong problems. The ideas of reproduction, metabolism, entropy, order, etc., while possibly relevant to your or my existence, do not instill in us the visceral sense of what it is to be alive in our day-to-day experience.

When asked what it is like to be alive, the most characteristic features we humans usually point to are ones related to the concepts of agency and free will. As technical terms these have defied ready definitions or explanations because right now they are based more on our experience of the world than what is captured by our scientific understanding of it.

Free will is a familiar subject to popular debate, so let's start there. You probably feel like you have free will. I do too. In fact, I feel like I chose to write this book, and you may feel like you chose to read it. Current popular accounts in physics would claim this cannot be true. For example, Brian Greene, the acclaimed science communicator and string theorist, writes, "The processes of life are molecular meanderings, fully described by physical laws that simultaneously tell a high level of information-based story."[18] I first came across this quote from his recent book, scrolling through posts on Twitter (now X) while cooking breakfast. What stopped me cold on both fronts (cooking and scrolling, which can be risky when done simultaneously) was not the posting of this quote by a fan of Brian's book. Instead, it was Brian's own response. He tweeted back, "Reducing life/mind to its molecular basis does not in any way diminish life or mind, but rather aggrandizes both: look at the amazing and wonderful things particles can accomplish." I had seen videos of Brian arguing how physics has no need for free will, and therefore neither should we: current physics provides an elegant enough description of the universe. But could anyone really believe physics at this moment in history provides an ultimate explanation for what and why we are?

Let's take a step back to illuminate the conflict between free will and current physics. Free will is generally considered the capacity to act at one's own discretion, independent of your current state or history. The laws of physics, at least as we understand them now, describe a universe that is fully determined from the beginning. Everything that happens is literally unfolded in the dynamics of elementary particles and fields. Your thoughts and feelings have no impact on reality, let alone you. There's no room for free will, because everything about you was already determined in the initial state of the universe. End of story. Or is it?

Many scientists assume that all causation occurs at the most microphysical layers of reality, at the fundamental scale of elementary particles. This leaves no room at large scales for what is called higher-level causation, e.g., for us to matter to the universe at all. You may think that your thoughts control your behavior, but the counterargument runs just as Brian declared it: those thoughts are really "just" informational patterns that emerge from the interaction of thousands of neurons, each made up of thousands of molecules interacting, and each molecule is a collection of atoms that are in turn composed of elementary particles—all subject to (known) physics. In this way, a thought—which is a pattern carried by particular chemical and neuronal arrangements—can be reduced to its most basic causes at the level of elementary particles. If the laws of physics have sufficient explanatory power to cover what happens for the low-level things reality is made of, maybe that really is enough to explain it all. The rest is just a constrained outcome within those laws.

Daniel Dennett, the prominent philosopher of mind, has a telling example of the tension this presents in his essay "Herding Cats and Free Will Inflation,"[19] in which he makes the important distinction between causation and control. He uses an example centering on the contrast between a boulder and an expert skier hurtling down the same mountain. The boulder, as Dan says, is "out of control"—its trajectory is determined by the laws of physics, but it is not controlled. The skier's trajectory is also determined by the laws of physics, but she is in control. Her decisions, her skill and strength, the conditions of her skis: all go into determining her trajectory. Causation and control are not the same thing, Dan then argues, because all things are caused but not all things that are caused are controlled. Control requires an agent. Agents can control a process via feedback: information about the trajectory and conditions can be used by a controller to regulate its action. Agents gain self-control via evolution and learning, and this is also how they can control some aspects of themselves and their environment.

A focus on agency and control should be contrasted with "free will inflation," as Dan calls it: most arguments on free will assume it is all or nothing. The "all" gives free will in all situations, such that nothing can be said to be determined. This is what Brian argues against: physics (and indeed reality) is not a random playground for agents to make any decision we want. We observe things to have determined behavior, particularly at the microscale. The "nothing" removes free will as a possibility at all, because it assumes causation (without control) is sufficient to describe everything. This view ignores that agents also can influence what happens. Dan resolves this by pointing out that you cannot have control over all causation, but you can have it over some. It follows that sometimes you can exert free will, but not always, because you cannot control everything. Free will is something we have sometimes, but not always.

While Dan resolves how free will cannot be all or nothing by invoking agency as a causal category, there is still an open question: we do not know what gives agents this special causal power of control. Indeed, Brian intersects similar concepts in his idea of informational patterns but fails to give a mechanistic account of the physics of what those are.

In the 1970s, the physicist Philip (Phil) W. Anderson, who was awarded a Nobel Prize for his fundamental contributions to our understanding of matter, boldly declared "More Is Different" in a hugely influential essay in a leading scientific research journal *Science*.[20] Phil was among the most prominent voices opposing the reductionist perspective shared by Sean, Andy, Jack, Brian, and others I've already highlighted. Phil's central argument claimed that while we can reduce the components of the universe to elementary building blocks, and in turn describe these by reasonably simple laws (the edifice of the last four hundred years of physics), this does not imply that we can reconstruct the universe from those laws alone. Each new scale—moving up from elementary particles to atoms to chemistry to biology to technology—might have its own fundamental rules or laws, not fully reducible to the lower levels. His work inspired much of the complex systems science that followed. He was involved in the early days of the Santa Fe Institute (SFI), which has been described by David Kushner in *Rolling Stone* as a "Justice League for renegade geeks."[21] At SFI you will find the kinds of geniuses that like to push against standard paradigms, building the scientific understanding needed for the next century, not the past one. Indeed, thanks to SFI, we now have a sect of scientists that hang out in the desert Southwest of the United States, who think about complex things, and often are found proudly wearing the slogan "More Is Different" on their T-shirts.

It would probably have been a bit less interesting to declare "More Is the Same." I am glad I do not live in that universe, and that some of my colleagues share my sentiment that such a universe might be far more boring than the one we live in.

What physics explains right now, very deeply, are some fundamental aspects of reality but not nearly all of them. Yet modern physicists have been rather ambitious in their desire to explain everything. Some literally think we can have a "theory of everything." Such a theory would be able to explain the origins of gravity, the elementary particles, and by extension—or so some think—everything else, including the agents that Dan points to, if only we had a computer large enough to simulate them with known physics. But in the words of David Krakauer, president of SFI, "A theory of everything is a theory of everything except those things that theorize."[22] For all the talk of the explanatory power of a theory of everything, any attempts at making one are to date missing an important feature of our reality: us.

Personally, I am convinced there is entirely new physics in the living universe (the part of reality that includes living things), awaiting our discovery. Or at least I am convinced that adopting this view is more productive for gaining traction on hard problems like the origin of life or finding aliens than assuming there is nothing fundamentally new to discover in life. More radically, it is not just new laws in the old paradigm that we need to dream up, but new ways of doing physics and new implications for our understanding of seemingly elementary things we think we already have a handle on, like matter and time.

To get to an answer for "What is life?" we may need to ask a different question: "Why life?" That is, we must transition from *defining* life based on a set of descriptors possibly unique to our Earth, to *deriving* life's properties from a fundamental theory of the kind physicists are known for developing, which not only explains what but also why. This is necessary if we are to have any hope of solving how life first emerged on Earth or generalizing our understanding of the possibilities for life that may lie beyond the boundaries of our own planet.

I am alive and you are too. It's time we had a physics that could explain this, but we might need new explanations to do so.

Natural Kinds

When I was an undergraduate student, our physics professors would often tell tales of how shortsighted scientists were at the turn of the nineteenth century into the twentieth. "They thought all the problems were solved!" I would hear one professor chuckle in a course on classical mechanics. "There was nothing fundamental left to be discovered!" Another would gasp in a course on thermodynamics. "The prevailing thought of the day was that all that was left was technical refinement," to yet more amusement in a class on quantum mechanics. I would laugh too. So long as I studied what physics departments taught, it seemed that shortsighted views of the future were just a history lesson—one we were taught about the past but that we were not taught to look out for in the present. "Look at everything we did in the last century!" my professors would say, seemingly amused by how closed-minded their predecessors had been not to see quantum mechanics or general relativity coming.

It did not occur to me that one hundred years from now, future physicists might be saying a similar thing about the turn of the twenty-first century. That is, until I was hired as a professor intending to apply the approaches of theoretical physics in a new way to the problem of the origin of life.

With some trepidation I followed my then postdoctoral mentor, Paul Davies, director of the Beyond Center for Fundamental Concepts in Science at Arizona State University (ASU), to the office of the dean of the College of Liberal Arts and Sciences to discuss the details of my hire to the faculty at ASU. In the very academic environment of a disheveled office, with papers scattered on tables and unerased whiteboards on the walls, we discussed what school within ASU would be my most fitting tenure home.* The options were the School of Earth and Space Exploration, the School of Life Sciences, or the Department of Physics. "To be frank, it doesn't matter which you choose, they all have equally deplorable numbers of women faculty," the dean said, looking me straight in the eye. Paul raised an eyebrow. This did not faze me, and neither did the reality that only two of the stated three options were really available: the physics department already thought I wasn't hirable—the problems I worked on were too far outside of the scope of what physicists do, even though I was formally trained as a theoretical physicist and approached my work as such.

Lee Cronin, the Regius Professor of Chemistry at the Univer-

*I was approved for hire by the provost before a tenure department. For those of you who know how universities hire, you will know this is a bit backward. But I had a large grant and a competing offer from another institution, so ASU decided it might be worth keeping me.

sity of Glasgow, describes what our field is going through as "pre-paradigmatic": understanding what life is and how it arises is exciting because there are no established frameworks that researchers need hold ourselves to. For those of us in the trenches, this can be a source of inspiration: pre-paradigmatic science affords more room for creativity. It's also a challenge. You are hanging over the edge of what we know and need to raise money to do it, convince colleagues you are on the right track (or in the case of the problem of life, sometimes that there is a problem to be solved at all!), and convince students and postdoctoral researchers to risk their careers working on hard problems with you. Science is first and foremost a social endeavor.

Lee's reference to Thomas Kuhn's idea of science in its pre-paradigm phase is apt. Thomas Kuhn was a historian and philosopher of science who introduced the concept of a "paradigm shift" in his highly influential book, *The Structure of Scientific Revolutions*,[23] published in 1962. Thomas's key argument was that science is not continuous in its generation of new knowledge: instead, there are major and abrupt transitions—*paradigm shifts*—that open new approaches to understanding the world that scientists would not have considered valid in earlier generations. A key feature of his idea is that what constitutes scientific truth is informed not only by what might be objective features of the world, but also the consensus of the scientific community at a given point in history. Immature sciences, or those that are pre-paradigmatic, are those that lack consensus. This lack of consensus can take many forms: in the relevant questions to ask, in the answers we might find, or in who is qualified to ask and answer the questions.

There is no standard paradigm that defines a discipline squarely housing the study of what life is, nor are there standard methods of study for how we should approach this problem. The boundary between the phenomena we want to think of as life and not life is fuzzy at best and may not exist at all.

We cannot always see this clearly because of the arbitrary boundaries we set between the current classification of disciplines we think are needed to solve the problem, which are based on paradigms not suited for solving what life is. I am a physicist by training, so I must constantly check the biases I bring from that training into how I think about the world and my work on the origin of life.

In my experience, and broadly speaking, biologists approach the problem by defining life in terms of observed features of life on Earth, which is not especially useful when you're looking for life's origins or for life elsewhere in the universe. Astrobiologists need guiding principles to inform how they conduct their search, but they, too, end up being overly anthropocentric in their reasoning: their search is most often directed at signs of life that would indicate biology exactly as we observe it here on Earth. Chemists either think life does not exist or that it is all chemistry (probably these are equivalent views). Computer scientists tend to focus too much on the software—the information processing and replicative abilities of life—and not enough on the hardware, i.e., the fact that life is a physical system that emerges from chemistry, and that the properties of chemistry literally matter. Physicists tend to focus too much on the physical—life is about thermodynamics and flows of energy and matter—and miss the informational and evolutionary aspects that seem to be the most distinctive features of the things we want to call life. Philosophers focus too much on the need for a definition or the flaws of providing one, and not enough on how we can move as a community beyond the definitional phase into a new paradigm.

Nature does not share these boundaries between disciplines. They are artifacts of our human conception of nature, our need to classify things, and historical contingencies in how our understanding of the reality around us has evolved over the last few centuries. That is, they are the product of paradigms established in the past. We are in part pre-paradigmatic in understanding life as a general phenomenon in the universe because there is no defined discipline that can fully accommodate the intellectual discussion that needs to be had about what life is—at least not yet.

Now, it may be the case that life does not warrant a deep explanation in the manner that other physical phenomena do. I am biased by my training as a theoretical physicist and by my desire for deep explanations, and I must admit that there may well be no such fundamental explanation. Yet given how attempts to define life based on common sense criteria have so far failed, it may be our best shot. I also find it hard to believe that quantum physics and theories of gravity reveal deeper aspects about the fundamental nature of reality than life does: life is after all the only part of the universe we know of that can comprehend the rest, including writing down these theories of physics. Surely that warrants an explanation that is equally as deep as quantum physics or relativity, if not deeper.

In the Beyond Center for Fundamental Concepts in Science

where I work at ASU, we regularly hold workshops on boundary-pushing topics, with provocative titles like "Why the Quantum?," "Quantifying Complexity: Can It Be Done?," "Nature as Computation," "The Origins of Meaning," "Mathematics: Evolved or Eternal?," and "Infinite Turtles or Ground Truth?" After one such workshop I was walking back toward campus from the meeting venue with Frank Wilczek. Frank is the modern instantiation of the iconic physicist—when you think of a physicist, you probably have a Frank-like avatar in your mind. He won the Nobel Prize for work on quantum chromodynamics done in his early twenties, and he has contributed foundational ideas to many other areas of physics, from proposing axions as a dark matter candidate to more recent work on the idea of time crystals. I greatly admire Frank, but I don't agree with him much on things outside the standard canon of physics, including his views on life or self-reproduction. On this occasion, we were discussing the question of what life is. I explained some of my ideas, to which Frank responded to the effect of "Well that is not an intuitive view." I was a bit perplexed, as it seemed completely intuitive to me, even if somewhat unorthodox.

While I was very confused that an abstract idea about what life is wouldn't resonate with one of the world's great physicists, I was also struck by what Frank said next: "Maybe the problem is life is not a natural kind." I wasn't familiar with the term at the time, so I had to go look it up after our conversation. Philosophers refer to categories as kinds, and a kind is natural when it reflects the structure of the natural world and not our own subjective views as humans. I'm pretty sure Frank was implying that the way we talk about life is not a category the universe recognizes but is rather a human-derived construct. I could not agree with this point then. But I have come to think that this is probably right. As is the fact that life does not exist, at least not in the way we currently think it does. That something is missing from the modern foundations of physics from which we might explain, and in turn derive, life's key features is, however, a very real possibility.

Two

HARD PROBLEMS

Three great problems that plague scientists and philosophers alike are the origins of matter, the origins of life, and the origins of mind. These were pointed out to me as the key problems of existence by Marcelo Gleiser, who had spent most of his career on the first problem, contributing to our understanding of how matter was created in the very early history of our universe. Like many other more senior physics professors, he was content enough to start dabbling in some of the other great open problems later in his career. He had just started on the second problem around the time I began working with him, hence his suggestion to me that I pursue the origin of life as the topic of my dissertation research. "The conferences are in much better locations," he joked, but he also told me more seriously that "the job market would probably be better in astrobiology than in cosmology." I did not much care about either of these as motivators, but I did care about big questions and open mysteries to be solved.

But before I had my chance to ask the big questions, I was asked another pertinent question: "Are you sure you want to do that?" I had just been accepted to the physics and astronomy PhD program at Dartmouth College, an Ivy League university—not bad for a former community college student whose parents did not attend university, or so I imagined. Accepting the offer to go to graduate school for cosmology seemed like the obvious thing to do.

"Why not?" I asked.

"Well," my professor responded, "because cosmology is pretty hard."

But I thought that was the whole point. I wanted to study cosmology because it's hard.

Of course, there's a big difference between the kind of hard this professor thought a kindness to warn me about, and the kind of hard I intellectually crave. Open scientific problems can be technically hard or conceptually hard. Cosmology is known for being both.

The mathematical rigor that goes into predicting features of things like particles and fields or the large-scale structure of the universe is technically hard. I once had a homework assignment in a graduate-level course on General Relativity that included a typo in one of the equations. After spending hours completing forty-six handwritten pages of calculations, I finally gave up on trying to get to an answer. I can still recall the moment my fellow grad students and I realized the mistake in the original posing of the mathematical question, meaning it did not have a closed solution. . . . I could have been going on to infinity! As students training in a highly technical craft, we did not catch the mistake, but an expert probably would have. Technical science is hard because a lot of skill goes into it, and this requires experience. Once you learn the logic of the rules and how to apply them, it's not such a leap to apply those same tools to new problems.

While I enjoy the technically hard, I went to graduate school because I love and intellectually thrive in the conceptually hard. These are the kinds of problems Marcelo posed with his three great problems of existence. You must learn to solve technically hard problems in order to imagine solutions to conceptually hard problems. But problems that are conceptually hard are quite different because they require paradigm shifts and demand entire new ways of thinking—we cannot just apply the old rules to the new problems.

There are many conceptually hard problems in science. What is intelligence? Can we make conscious machines? How will the universe end? Why do we exist? Is immortality possible?

At this point you may be wondering why of these I chose to focus on the origin of life, as surely it is not the deepest of the deep questions. Given the current state of progress in artificial intelligence (AI), for example, would it not be more timely to focus on the nature of intelligence, or even consciousness?

I will borrow words from another collaborator and friend, Michael Lachmann, an external professor at the Santa Fe Institute, who would say in response, "We study the origin of life because it is the easiest of the hard problems." By calling the origin of life an "easy" hard problem, Michael is not implying that it will be easy to solve. It is a hard problem for a reason. What he means, which I intuitively agree with, is that it is the right question to be solved at this moment in history. Solving it will open new ways of thinking that start to unlock the other (perhaps harder) hard problems. It is the easiest of the hard problems because it is the one that we can solve in this generation. In fact, many of us working on it now feel it is precisely the hard problem we need to solve in order to understand the technological transitions we are currently living through: many of our technologies appear to be taking on a life of their own, extending our biological lineages into new substrates, right before our eyes.

The Hard Problem of Consciousness

Right now, I am experiencing a lot of color. I'm sitting in my backyard and the bougainvillea are blooming. They are immensely pink, so loaded with flowers you can hardly see any green. What I am experiencing—being surrounded by pink blooms, the desert breeze, the ephemeralness of the moment—neuroscientists and philosophers will call qualia. *Qualia* is the word we use to label and talk about what we all seem to feel in our mental states rendering something, or nothing, or anything at all: it is the felt quality of being.

It is a deep mystery why being is like anything, rather than nothing. Some might argue it would be simpler for nothing to exist, yet here we are. Qualia are front and center in one of the deepest and longest standing intellectual debates we humans have ever engaged in—the nature of our own minds.

Early in human history we realized others probably have experiences, too, that it's not just a fluke feature of one of our minds. By naming the things we experience as *qualia*, we can discuss the idea of it with one another. We can talk about the nature of intrinsic experience and what it feels like to be one of us. But we do so without ever knowing what it is like to be another experiencer. Indeed, by definition, we can never know what it is like to be someone else.

This is what makes the hard problem of consciousness so hard. This notion of a hard problem of consciousness was popularized by the philosopher of mind David Chalmers.[1] David intended to focus researchers on precisely articulating what current science cannot explain and perhaps cannot even attempt to explain about consciousness: subjective experience. Why should it feel like anything to exist at all? Surely things could exist and be exactly the same even if nothing had an inner world, as we seem to have in conscious experience. It seems there is no way to test scientifically the existence of subjective internal experience. How can we test what it is to be you from the inside, when we will always, by definition, be outside of you in any scientific measurement we can perform?

Over the last few decades, scientific studies of consciousness have primarily focused on what are often called the easy problems of consciousness, aiming to locate "neural correlates of consciousness." These are things we can measure that map to reports of what we experience. They are technically "easy" because solving them seems within reach with current science, and there is a general understanding of the steps needed for the solution to become apparent. In looking for neural correlates, neuroscientists are specifically aiming to identify features of the brain that map to specific experiences an individual might report, i.e., a network of minimal neurons that might be associated with an experience. Neural correlates might be measured using technology like electroencephalograms (EEG) or functional magnetic resonance imaging (fMRI), which can assess brain activity during different behaviors.

But even if we do succeed in eventually uncovering a complete mechanistic understanding of the wiring and firing of every neuron in your brain, it could very well tell us nothing about your thoughts, your feelings, and what it is like to be you experiencing something.

The issue forces us to reconsider much about what we know about physics. David's original proposal for a resolution to the hard problem of consciousness was to regard subjective experience as an irreducible, fundamental property, with its own laws that cannot, even in principle, be reduced to the known laws of physics. If you agree with this view, then your logical conclusion should be that the most progress to be made by scientists attempting to understand consciousness is in the development of a testable theory of consciousness to stand alongside our theories for matter.

To understand consciousness, scientists aim to measure the subjective, defined as a physical system's internal or intrinsic experience of the world. However, we cannot do this because we can study only objective features of its measurable properties. This is the reason that the hard problem is hard, and possibly even outside the scope of science (at least as we've understood how to do it so far). But is this problem really unique to the study of consciousness?

The Hard Problem of Matter

I cannot know any more of what it is like to be an electron than I can of what it is like to be you. Perhaps science cannot resolve this. Personally I think I have a better idea of what it is like to be you than what it is to be an electron. Even the idea that it "is like" something to be an electron feels hard to comprehend. But experiencing electrons are an important consequence of one of the proposed resolutions to the hard problem of consciousness: perhaps consciousness is fundamental, and therefore all matter is conscious.

Defining matter is usually in the purview of physics, not neuroscience. An example is the very electrons in question. Electrons are elementary particles defined as indivisible units with a very small but measurable mass, a negative charge, and a spin (a measure of a particle's internal angular momentum) of +½ or -½. These are properties we associate with an object we have confirmed to high confidence exists in our universe, which we call the electron.

You might naïvely think that the electron really does have a mass, a charge, and a spin, and that these are intrinsic properties of the electron. But these properties can also be considered to merely describe how electrons interact with certain measuring devices. For example, when an electron interacts with scientific instruments that measure mass, they provide a reliable value we identify as the electron mass. When these same objects interact with devices that measure charge, they reliably report a negative charge. Spin can be inferred in a number of ways, including by measuring magnetic properties of particles or their interactions with other particles. We have no way to know what other properties electrons may have that we haven't measured or been able to infer by measurement of their interactions with other objects. Nor can we know if electrons still have the properties we know about when they are not being measured. For practical purposes it may be perfectly reasonable to assume they do, but we can never completely confirm it.

In fact, we will never have complete information about any object that exists in our universe, because we can never know about an object without it interacting with another (e.g., at some point the interaction must involve a measuring device). We do not know what it is for any object to just be. This is the hard problem of matter, first coined by the philosopher Galen Strawson in a *New York Times* article.[2] It points to how we are challenged to understand what is physical, because understanding anything that exists outside of measurement is deeply problematic.

Among Marcelo's hard problems, the origin of matter is considered solved by physicists, because we have a pretty good understanding of baryogenesis, the process that biased the formation of matter over antimatter in the very early universe. Yet we still do not have a concrete answer to the philosophical question, What is matter anyway?

Defining consciousness in some ways shares the same stubborn property as defining matter—it can be done only from the outside. An easy way to kill two hard problems with one stone is to make the unexplained thing fundamental. What I mean is, if you must take something to be true, e.g., that consciousness or matter exists, even though you cannot prove how or why, it might be easier for subsequent reasoning to just assume they are the same thing. Whatever it is that makes consciousness material is fundamental; you can derive the rest of your reasoning from there. Mathematicians do this all the time when they declare axioms: examples include defining points exist, or that it is possible to draw a straight line between any two points. While these seem self-evidently true, and we can derive nearly all of Euclidean geometry from such simple statements, it could also be otherwise. For example, if we relax some of the axioms of Euclidean geometry (something that took humans thousands of years to realize we could do), you end up with self-consistent geometries that have very different properties: these, for example, became important in describing the curved geometry in the physics of space-time, which is not Euclidean. It is important to keep in mind as a theory builder that whatever you assume as the axioms in your mathematics, or fundamental in your theories of physics, constrains and in many ways determines your subsequent reasoning and worldview. It defines the boundary of what you can explain.

Let's adopt the view that experience is the default of existence. That is, we will assume experience is the material things are made of, the stuff you cannot observe from the outside, and see where it takes us.

The first conclusion is that if this is true, then electrons must be conscious too.

The philosophical belief that all matter is conscious is popularly known as panpsychism. Panpsychism purports that the physical structure of all material objects is literally made out of conscious experience. This inverts the usual logic. Usually we assume physics deals with the hardware of reality—the physical structure, including material objects and their interactions. Consciousness must then describe something more akin to software because it is more intangible, abstract, and deals with the inner workings of objects we cannot measure via interaction. But as the philosopher Hedda Hassel Mørch points out,[3] in the panpsychist conception of nature the resolution is to say consciousness is the hardware, the stuff that really exists in every object. The interactions physicists write in mathematical form are then merely a description of the software characterizing only those subsets of features we can observe and measure via interaction.

Even if you do accept the premise of panpsychism, you are still left to explain what consciousness even is, and also why and how it is that our experiences are presumably not the same as those of electrons. Or maybe they are? This might be considered easier than the original hard problem, but it is not easy. As science writer Annaka Harris points out, panpsychism does not fully capture what is needed, to do that might demand a new idea and even new physics.[4] Currently, no viable theories to empirically test the validity of scientific ideas deriving from the philosophy of panpsychism have been proposed.

Both the hard problem of consciousness and that of matter seem intractable to answer in our current scientific framework. Usually the best way to tackle hard problems is to play with them until you find a window that opens up new questions and experiments. In this spirit, maybe the resolution is indeed to combine these dual hard problems as the panpsychists aim to do, but to do so in a way that does not assume consciousness is the substrate of all objects. That is, let's not assume consciousness is fundamental, but instead that something is missing that can explain what it is. If consciousness were the most fundamental thing (our first axiom), it leaves us with no room to derive or explain the features of consciousness, or to test features it might impart on some objects and not others. Recall, we are after explanations with measurable consequences here, because we are aiming to prove new theories that explain reality within a scientific worldview.

Instead can we ask, if consciousness is a material property of at least some objects, what does it do or cause in the world that we might observe only for those objects that have it? Does this allow us to see something deeper that explains *what* consciousness is and *why* it exists?

This is a very different sort of question. Instead of asking what consciousness is, we ask: What does consciousness do? Recall our motivation in this book is developing testable theories for what life is. In order to test a theory, the contents of that theory—the things it abstracts—must have measurable consequences on the world. They must cause things to happen. By focusing on whether consciousness causes anything we could not explain otherwise, we re-center the problem of consciousness on something measurable. But here we are not measuring whether something has experience (what consciousness is), but instead whether something that has experience can do different things because it has an internal world (what consciousness does).

Daniel Dennett proposed this alternative to the "hard problem" of consciousness—he calls it the "hard question" of consciousness.[5] His aim was to refocus his colleagues' efforts. If consciousness does have some measurable impact in the world, might we target identifying that and test our ideas via the well-trodden methods of science?

If consciousness is a real thing, and not just an illusion or epiphenomenon, as some people will claim, then it should have some measurable impact on how a physical system behaves. That is, if consciousness is physical, or even a natural kind, its presence should have dynamical consequences. So far, much of the science of consciousness has focused on what happens within an individual brain. But it may be that what consciousness does—as far as the causal or dynamical consequences it has on our reality—is not testable at the level of the individual mind. It could be that the physics of consciousness can be measured only in collections of minds. Only by sharing features of our individual subjective experiences with one another (each subjective experience inaccessible from the outside) can we generate something genuinely new in the universe. So we need to ask, does anything in the universe exist that might not be possible if subjective inner worlds did not exist?

The answer, I think, is yes. There are things like rockets, which existed within our collective imagination for centuries before they were ever built as physical objects with definite properties. There is nothing about rockets that violates the laws of physics in our universe. But at the same time, rockets do not form spontaneously anywhere in the universe, because they require specific knowledge to build them. That rockets can be made to happen, once minds emerge that can imagine them, is a nontrivial feature of our universe. Rockets are physical evidence of imagination. Rockets are not special in this regard; many things in the human environment are evidence of the inner world of the human mind: computers, cars, houses, gardens, fashion, etc.

Consciousness may be evident when new things are so statistically improbable that a new abstraction, or idea that suggests those things are indeed possible and how they are possible, has to be invented by our universe to generate them. The idea has to pass through many minds to come into existence as a distinct physical object, separable from the minds and technology that generate it. If this is an accurate description of the physics of consciousness, could we formalize this intuition and identify more generally the physical markers of imagination?

The idea that the observational features of consciousness are collective did not come from my individual mind. I first found the idea in conversation with Takashi Ikegami, an often radical thinker (in the best way) working on artificial life in Japan. As Takashi says, "Artificial life is bigger than biological life," meaning we need to dramatically expand what we think life is to build artificial examples. In this particular conversation, Takashi proposed consciousness might be contagious, like a virus. His conjecture is that babies first learn to be conscious from their parents, and therefore the way to build conscious machines would be to teach consciousness to them, like a parent teaches it to a child. Others also argue consciousness is a property of the interaction between two physical systems. Lex Fridman, the AI scientist and podcaster, thinks that if we have the same experience of interacting with other living creatures when we interact with robots, that is good enough. It doesn't matter if the robot is sharing the same experience for the experience to exist. Our interactions with robots can be conscious, even if we are not both always conscious independently. By extension, perhaps the interaction of two robots could be conscious, too, but with a different quality to experience than our own.

At this moment in history, there are many things we can imagine that don't currently exist—just pick any technology we haven't quite built yet, like 3-D chemical printers deployable in regions without access to modern medicine, a lunar colony, or brain-machine interfaces in every home. Or imagine a perfect circle, which is an object that will never exactly exist (a perfect circle requires infinite precision to specify, which cannot happen in a finite universe in finite time). Nonetheless, your idea of a perfect circle could cause things to come into existence, for example, if you decided you wanted to 3-D print a circle that was as perfect as possible. In this case, your ideation of a perfect circle would be causal to the physical manifestation of an object that is a close approximation to your idea of it.

Taking the preceding into account, my conjecture is as follows: Some things that exist are imagined through abstraction (are counterfactual) and become physical (made actual) through a phenomenon deeply connected to what we call consciousness. It is not that all matter is conscious, but that consciousness is potentially a window into the mechanism for bringing specific configurations of matter into existence across time. Consciousness and matter are therefore deeply connected, and deeply connected to how we exist in time, because our consciousness sets boundaries on what we can make in the future: matter is what exists now, and consciousness is our evidence of the physical boundary of what might exist—it defines what we, as features of our universe, might build in the future. If this conjecture is true, consciousness creates the possibility for things to exist that otherwise couldn't because they did not exist in the past. It is what allows the counterfactual (what could have been or could be) to become the actual.

In the words of Albert Einstein, attributed to an interview with *The Saturday Evening Post* in 1929:

> I am enough of the artist to draw freely upon my imagination. Imagination is more important than knowledge. Knowledge is limited. Imagination encircles the world.[6]

The question is, How can we formalize and test such imaginative and abstract ideas and make them physical? To start to get there we need to visit our third hard problem: that of life.

The Hard Problem of Life

Most people think consciousness is the hardest of the hard problems, and I can agree to a point, but conscious experience also has the advantage of being directly perceptible to us. While you and I regularly experience what it is to be conscious (by definition), we never directly feel what it is to be alive—life, like everything else, is filtered through our conscious experience of it.

There is no consensus on what features of life are universal. In fact, as we saw in chapter 1, there is no general agreement that life even exists or that there is a problem to be solved. Those who argue life does not exist also do not agree on what it is they are claiming the nonexistence of. The concept of life is too poorly defined.

It is assumed by many active researchers that focusing strictly on synthesis of the known molecular building blocks of life will be sufficient to get to an explanation of life and its origin. For discussion purposes, let's call these the "chemical correlates of life," in direct analogy to the "neural correlates of consciousness." They are correlates because they are molecules we find in known life, but we do not know if they have properties universal to all life, and if so, what those properties would be.

The neuroscience community recognizes that identifying neural correlates of consciousness will not fully answer the hard problem of consciousness. However, many believe that chemical correlates of life are sufficient to solve the problem of the origin of life, and to identify alien life on other worlds.

Also in direct analogy to the neuroscience of consciousness, it is clear that most research on the origins of life focuses on "easy problems." These may be technically challenging to address, but at least we can envisage a way to do it. These include understanding the mechanisms underlying replication, compartmentalization, and metabolism—some of the key chemical functions of known life that commonly come up in the many failing attempts to define life. Compartmentalization provides a concrete example: life as we know it exists in cellular compartments. Mix hydrophobic lipids in water and they will self-assemble into lipid bilayers, often forming circular droplets with an inside and an outside that look much like cells. These are "lifelike," but are they life?

Focusing on chemical correlates has yielded limited progress in solving the origin of life, even after many decades of effort. This begs the question: Will all properties of life be brought under the easy category, and allow us to solve the origin of life? Or are we missing something more fundamental?

If you know me at all by this juncture in our discussion you will already know my answer: we are missing something. It is not just that we need new principles, but we also need to rethink how we write down laws of physics in the first place. We need new physics.

If we can narrow in on the key problem of life it could give us a window into what a theory of physics that can explain life will look like. With this in mind, my colleague Paul Davies and I aimed to identify which aspects of life will prove too stubborn in our attempts for reduction to known physics and chemistry. That is, we wanted to identify the property that constitutes a candidate for the hard problem of life.[7]

What we came up with, stated most bluntly, directs attention to the question: *How is it that information can cause things?*

Information is a challenging concept because it is abstract, and it is usually discussed in a handwavy fashion (I am guilty of that myself and will do some of that handwaving purposefully in a moment). The challenge is how to make the concept of information physical, such that we can recognize how it matters to matter.

My examples in this book will mostly be anthropocentric. Most books about life and its origins focus on nonhuman, biological examples. But in explaining life, my real goal is to explain us; that is, I want to explain all life on Earth, and humans are the part with whom we are most intimately familiar. However, if you prefer a more traditional example of information and why it is viewed as fundamental to life, genomes are nearly everyone's go-to. So let's start there before transitioning to human-centric examples, which I use because they make much more obviously apparent the same critical role information plays across all life, but in a place we can readily talk about the most critical features because we directly experience them.

Genomes carry information much like a string of characters in human language—the sequences of A, G, C, and T can be copied from one strand of DNA to a newly assembled strand in almost identical sequence (there are always a few errors). This information can also be copied onto a page of text (e.g., if I wrote out all the As, Gs, Cs, and Ts on paper). Even when copied from chemistry to human text it can retain the same meaning: our technology allows us to read out sequences from computers to synthesize new copies of genes. In a famous experiment performed by researchers at the J. Craig Venter Institute, the genome from one species of microorganism was transplanted into the cell of another species. They sequenced the genome, stored the information on a computer, synthesized the genes, and put them in a new cell. When booted up, the transplanted genome reprogrammed the host cell, converting the host species' cellular properties to that of the foreign DNA's phenotype.[8] Not only can the information in a genome be copied, it can cause the same things to happen in a different point in space and time given suitable conditions (e.g., a machine or organism capable of reading and executing the information in a manner consistent with its historical use).

There are also plenty of examples of nongenomic information doing interesting things across biology. Consider planaria, flatworms with their eyes on top of their head, studied in the lab of my collaborator Michael Levin, who excels at discovering nongenomic information-processing modes in biological systems. Mike's lab has shown how planaria can change their shape if you manipulate their ion channels but leave their genetic information intact. Mike and his team can make the worms take on a two-headed phenotype or a two-tailed phenotype (or even more exotic forms).[9] These changes are *heritable.* As Mike describes it, the information for shape is not explicitly encoded in the genome, but instead is stored as a bioelectric pattern across the planaria tissue, that then exerts top-down control of how the cells organize into particular morphologies. Here, it is clear there is a distributed informational system—hypothesized potentially as a bioelectric code[10]—that is "calling the shots." Planaria are regenerative organisms; if you cut them in two, each piece, as long as it contains enough cells, can reconstruct an entirely new worm. This is one of the ways planaria reproduce asexually in the wild—literally by tearing themselves in two and with each half regenerating to form a new worm.

I could go on about many examples from across biology but would like to move more quickly to the heart of the matter and where the real deep questions lie. So let's now consider my favorite example of a physical process that can happen only if information exists to make it possible: anti-accretion. Right now, our planet is anti-accreting matter into space. Anti-accretion requires a technosphere (that is, global-scale technology—in this case, us) that has, through its evolutionary history, acquired knowledge of the laws of gravitation and now can harness that knowledge to launch little metal boxes, or red sports cars, into space (some will argue whether it takes more or less intelligence to do the latter). Here, *knowledge* of the regularities we associate to gravity—that what goes up must come down, etc.—is the "information" I am referring to that plays a causal role.

To understand how this is different from standard physics, we can consider the more normally discussed physical process of accretion. Accretion is the process of planetary formation, where matter starts to clump together to form small rocky bodies called planetesimals. These become increasingly circular as more matter is accreted onto them due to their self-gravity, eventually forming planets if enough material gets stuck together. Accretion happens everywhere in the universe where stars form protoplanetary debris disks that are set in motion and can aggregate into clumps. It does not require any memory, or information accumulated over evolutionary timescales, to occur.*

Anti-accretion, by contrast, occurs on only one planet we know of as I write. That planet is the one we live on, and it is happening here because of us. Ejecting material into space is of course not forbidden by the laws of physics. A one-off ejection event can happen anytime, for example, if a meteor hit the surface of the Earth and ejected debris into space. But the type of anti-accretion I'm talking about is repeatable in a programmable sense: it is a reliable process in which we can reset the input and it can happen again. We can launch satellites to space every day if we choose, so long as resources are available. Anti-accretion is consistent with the laws of physics because it is not forbidden by them, but it is not explained by them either. It does not happen due to random chance.

Now you may be asking: How can we distinguish this type of reliable artificial event from the naturally occurring types that also happen repeatedly? On large enough timescales, wouldn't meteor strikes also be considered reliable and repeatable? Yes, they are. But this is precisely the point: what life does is turn something that

---

*You could argue it does require evolution in the sense that the first stars did not form planets around them because there were not enough heavy elements. Planetary formation is historically contingent in this way. However, this is a trivial sense of contingency, because given the heavier elements and a protoplanetary debris disk, we expect planets to form readily from gravitational physics. There is no evidence for a nonzero likelihood of satellites spontaneously forming from planetary geochemistry and flying off into space in the absence of the evolution of a biosphere into a technosphere that learns something about the laws of gravity.

can happen, but perhaps is so exponentially rare or impossible as to never be observed, into something that can happen with very regular occurrence and be observed in abundance. Perhaps you are not surprised by a single event of ejecta that start orbiting a planet, but if you saw two, or three, or ten thousand in a short interval with precise control, your degree of surprise would increase. Physicists will sometimes talk about the laws of physics as universal regularities in our universe. What life does is create local regularities—these are things that can happen repeatedly with high likelihood, but only in a local part of the universe because they require memory and information stored locally to happen, and these can be acquired only via evolution.

Let's go straight to the heart of the matter: Why is Earth special in this example? And let's ask like a physicist would.

For anti-accretion as I've described it (repeatable, high likelihood launching of satellites to space) to be a possible process, a planet must have evolved a biosphere that has through its evolutionary history acquired the knowledge necessary to launch objects from its planetary surface into space in a reliable way. David Grinspoon, a planetary scientist and astrobiologist, notes that this is a key feature of the Anthropocene. The Anthropocene is what scientists have named the modern geological epoch, where it is now abundantly clear humankind has made an indelible impact on the Earth system. Our presence will be preserved in the rock record for eons to come, even if we do not survive so long. David has a nice vision of the precise moment the Anthropocene began: to his mind, it is when Neil Armstrong put the first human footprint on the Moon[11]—a geological artifact that will stand the test of time unless a fluke meteor impacts the surface at that location, or a teenager not yet born decides to play a joke on a future-day historical Moon museum. Importantly, the geological artifact of the lunar footprint as a predictable feature of our universe traces the lineage of its causation all the way back to the origin of life on Earth. If you start after the origin of life, you might have anticipated it could happen, with increasing predictive power the closer to modern times you are. But if you checked Earth just before life emerged, you could not predict it.

The geological record suggests life first emerged on this planet more than 3.7 billion years ago. Over the subsequent several billion years, cells evolved into multicellular organisms and eventually to creatures like us. Having evolved some 300,000 years ago and inventing agriculture approximately 10,000 years ago, leading to our scientific revolution 500 years ago, we have come to understand our universe in a rather deep way. We (humans and our technology) understand that there is a regularity—a predictable and recurrent feature of what we observe—associated with how objects move, and that what goes up must come down. Over time we turned this regularity into a law describing how it is that massive objects are mutually attracted to each other, via a force we call gravity. That law is formalized mathematically and allows precise predictions of the behavior of two gravitationally interacting objects, such as our Earth and our Moon, or our Earth and you. It also allows us to calculate how much counterforce is necessary to escape Earth's gravity, e.g., by accelerating an object to its escape velocity.

I am emphasizing this example to highlight how the laws of physics we write down are themselves an example of information. To explain what humans do when we write down laws of physics, we have to write new laws of physics that include us as part of the system we are studying. We have to generate new knowledge that includes an explanation of what knowledge is.

Here I'm defining information (and knowledge) as things that can be copied and retain their causal power. Asking what physics describes the laws of physics is the same as asking what physics governs information. We will get to unpacking more of what this means shortly.

The words I am writing right now are information. They started in one physical substrate (presumably the wet chemistry of my brain) and were typed out by my hands onto another physical substrate (my computer), where they were stored for many months in electronic form. They may get to you in a printed book or as an audio book, before finally arriving as (presumably) mostly the same information in the wet chemistry of your brain. *This is a key feature of information—it can be copied between very different things.* This is why we call information "abstract": it seems not to depend on the physics of the substrates you store it in or access it from.

There is another key feature of information I can demonstrate for you with an experiment. I'm going to ask you to do something, and for the sake of a successful experiment, I hope you do it (I am a theorist, so this might not work). Here's the experiment:

Turn the page.

This page is the experiment. Turn the page again.

We are separated in space, time, and in physical media, yet information can be reliably transmitted between us. It can cause things to happen—like you turning the page of this book and recognizing that I am playing a bit with the words in it, their meanings, and therefore what causation they carry.

Once acquired, in principle you can do the same things with information as I or anyone else can—I would be interested to know how many pages in how many books were turned due to the above prompt. This is the second key feature of information—*it causes things to happen.*

Human-invented mathematics, human-invented language, the information content of genomes and bioelectric patterning in tissues behave in much the same way when you abstract them to such a level (as do many other things in the biosphere and technosphere; maybe you can think of other examples). This universality suggests a broad regularity in information that might point to underlying universal laws.

We can gain more clarity on the regularities of information by returning to our discussion of the mathematical laws of physics as themselves physical objects (in the form of knowledge) that cause things in the world. Mathematical statements can be copied reliably between different things. Math is an even more exact window into the regularities of nature than what we see in genomes or in human languages. When math is copied between things it retains more of its original properties. I can tell you right now the sky is blue, and the flowers I am looking at are pink, but you will not have an exact image in your mind about what shade of blue or pink, or what kind of flowers I am looking at. This is because language is imprecise, and the causation in words is subject to a large degree of variation across minds. But if I tell you F = ma, and you know what the symbols mean, you will know that the force on an object is equivalent to its mass times its acceleration. You could then do an experiment and confirm it. You could use it to design your catapult for a pumpkin-launching festival. You could launch a satellite to space. You could do any variety of things with it because the information is accurately preserved when it is copied to move between minds and machines. If you tell your computer the same information, by writing it in a programming language (or now in symbolic language if you are conversing with a large language model), your computer knows it, too, and in the exact way that you do.* What the information can do in the world is retained (although what it does do in actuality may be radically different).

The concept of information I am adopting here is very much inspired by the work of physicists David Deutsch and Chiara Marletto on constructor theory.[12] When I set out to study the origin of life with Paul, we were part of a collaborative network of researchers including David and Chiara, Michael Levin, the philosophers Paul Griffiths and Karola Stotz, the neuroscientist Giulio Tononi, physicist David Wolpert, and others interested in the causal power of information. Constructor theory's unique solution is to describe what is possible and why. As Chiara puts it, we need a science

---

*This point may be confusing for the reader, and I should clarify. I do not mean the computer understands the information the same way you do. Indeed, I would argue the opposite. What I do mean is that the information it carries has the same meaning as an external-facing attribute: that information can, in principle, cause the same actions in the world. Thus, the feature of relevance here is not the subjective experience of the information, but the objective impact it has on what can happen.

of what "can and can't happen."[13] Chiara, David, and their colleagues are intrepidly working to rebuild physics from the ground up with this new perspective in mind. In many ways, constructor theory is not so much a theory of physics as it is a theory about physics. In constructor theory, the only transformations that can be caused to occur are those for which there exists a constructor. An example of a constructor is a chemical catalyst: there are some reactions that would be so unlikely as to never happen unless there is a catalyst present that causes the molecular transformation to occur. The concept of constructors in constructor theory is much more general than this and would apply to more abstract things like knowledge—for example, the laws of physics are a kind of knowledge that exists in our biosphere and that are part of the constructor that causes the transformation of launching satellites to space. Later in the book, we will discuss further the idea of constructors, which takes intellectual root from the work of physicist John von Neumann in his idea of a universal constructor. Importantly, constructor theory does not distinguish between things we consider as material, like a catalyst, or abstract, like knowledge, as being causes for what can occur in our universe. If correct, the principles of constructor theory should supervene on the laws of physics, including the laws we have identified to date and any laws we might discover in the future. The insight David and Chiara bring with their approach is that the informally conceived notion of information that we all have—i.e., that it can be copied and retain its properties, even in radically different physical materials[14]—means certain interactions are possible in nature and others are impossible. That is, it implies the existence of certain regularities in our universe.

Universal regularities—like how all massive objects attract one another—are most amenable to being cast as laws or principles in physics. A fundamental physical theory of information that explains life would therefore be one that has an explanation for the very broadest regularities that extend across all examples of life, from biochemistry to technology. We do not know what that physics will look like yet, but we are almost ready in this book to introduce some features of what it might look like.

The Hardest Problem: What Exists and Why

We have encountered three hard problems now. The *hard problem of consciousness*: that existing feels like something (at least for us). The *hard problem of matter*: that nothing can be observed to exist outside of interactions. The *hard problem of life*: that abstractions (information) matter in determining what can exist. These all deal in some way with the nature of existence: *what it is to exist (consciousness), how we determine what things exist (matter), and how some things can be caused to exist, and others cannot (life).*

Cast in this way, all three hard problems become one more fundamental problem we cannot seem to avoid any more than we can seem to answer it: *Why do some things exist (or experience existence) and not others?*

It is perhaps the most perplexing question of our existence that anything should exist at all. And if something exists, then why not everything?

Imagine the space of all possible pieces of art. Why do we paint, sculpt, laser, or blow into existence some of these and not others?

Imagine the space of all possible pieces of music. Why do some genres exist on our planet and not others? Imagine all tablelike objects, all chairlike objects, all catlike entities, all treelike entities, all cloudlike things, all possible elementary particles, all laws of physics: Who or what decides which take physical form to make up our universe?

One of Paul's talents as a mentor is to frequently pose deceptively simple, conceptually deep puzzles. One he posed to me early in my career was the problem of what exists.[15] Imagine all the things that could exist are in a great urn. You can choose to draw things from this urn, but the number of possible things is always going to be exponentially larger than any sample you might pull out. Paul's question was: What determines what is selected from the urn? What things get to exist, and why? Some imagine a being who literally decides. You might call that being God or something else, depending on your beliefs or training. If you are an atheist or agnostic, or a scientist who categorizes the mysterious simply as the yet-to-be-demystified, you might ignore the question entirely and assume existence the default.

The problem raises the question of why our universe is as it is. A different draw from an infinite urn of possibilities could yield entirely different laws of physics and thereby entirely different universes. As I write, some of the biggest questions in the physics community pertain to whether there is a theory of everything—one that might unify our theories of quantum physics and gravity (yet still not explain us, as I pointed out in chapter 1). But, in the words of Stephen Hawking, the famous cosmologist and public intellectual, in his breakout bestseller *A Brief History of Time*:

> Even if there is only one possible unified theory, it is just a set of rules and equations. What is it that breathes fire into the equations and makes a universe for them to describe?[16]

Mathematics is theoretically infinite, so determining why some math corresponds to our universe and the rest does not is a highly nontrivial problem. In fact, some argue that the set of possibilities for mathematical descriptions of a universe might really be infinite, not meaning this as a metaphor or a theoretical idea but a physical one. This is the view of Max Tegmark, a physicist at MIT who started his career in cosmology and moved into AI, and who is sometimes called Mad Max for what some regard as his unorthodox views. I respect Max's unorthodoxy in many cases, even as I disagree with some conclusions he draws from it. Max's proposed resolution to the problem of why some mathematical objects correspond to reality and exist as laws, and some don't, is to say that every mathematical object is physical somewhere in a multiverse of possible universes. This is the foundation of a proposed theory of everything called the Mathematical Universe Hypothesis (MUH).[17] In effect, the MUH claims that math and physics are the same thing: reality is math, and all math exists as physical. Universes that have entities like us that are aware will then be a restricted set of the mathematical multiverse, because our existence requires very specific kinds of self-referential relations between mathematical objects. Max's proposal is both fun and thought provoking, with a certain aesthetic elegance to it.

However, I've also found the MUH deeply unsatisfactory because it attempts to explain why the things we see exist by saying all things exist somewhere. Saying some things exist because everything exists ends up explaining nothing specific about why we see what we actually do see in our universe. It is surprisingly difficult to build a theory that explains what we observe and *only* what we observe to exist.

Many people, like Max, assume a Platonic reality for math: that all mathematics exists outside the human mind in an otherworldly perfect realm. Culturally, we teach math as something unphysical that was not generated within our universe. This makes it hard to directly argue against, because we all carry a culturally engrained implicit understanding of what we think math is.

So, instead of math, let's start with a physical example more obviously constrained never to exist all at once: molecules.

In fact, the easiest place to ask rigorous questions about what can and cannot exist is in chemistry. The stable and abundant elementary particles—protons, neutrons, electrons—do not have very much variation, and it is relatively easy for the universe to make them, given the right conditions and energy. These elementary particles come together to form the atomic elements. Humans have studied 118 of these in the lab—element 118, oganesson, is made possible on this planet because there are intelligent beings with knowledge of nuclear physics who can synthesize and stably store it long enough to measure its properties. It is not produced "naturally." In this way, oganesson is much like the satellites example in the last section—oganesson is possible because the laws of physics in our universe do not forbid it, but its existence in abundance requires information beyond that specified by the currently known laws (this time I am referring to our knowledge of the laws of nuclear physics).

More familiar, everyday examples of elements include carbon, nitrogen, hydrogen, oxygen, phosphorus, and sulfur. In fact, although there are 118 confirmed elements, most of life's chemistry is made of just these six. Chemistry is what happens when elements are combined into molecules. Currently, the largest chemical databases in the world include lists of tens of millions of molecules. That sounds like a lot of cataloged molecules, but it pales in comparison to estimates of the number of possible molecules. For example, cheminformaticians estimate 10^{60} possible pharmacologically relevant small molecules made with just C, H, N, O, and S atoms with <30 atoms per molecule.[18] It is important to note this estimate is for small molecules only and for a very restricted set of elements: no one knows the true size of chemical space. Because the space is so large, not all molecules can exist—there are simply not enough resources available in the observable universe to make all possible molecules. Unlike elementary particles and atoms, molecules are special because they are the simplest combinatorically built matter where not all the things that could be built will *ever* exist. There is not enough material, energy, or time to build them all.

In 2015, about one hundred scientists and philosophers met at the Carnegie Institution for Science headquarters in Washington, D.C. The meeting's topic was "Re-Conceptualizing the Origins of Life," and I had been pulled in to chair the meeting by a motley group of colleagues. Our aim was to shift the focus of approaches to the origins of life from traditional prebiotic chemistry to how the combination of theoretical advances, artificial life, messy chemistry, and new ideas about information and matter could energize progress.

Against the backdrop of the historic Elihu Root Auditorium, with its majestic murals of lunar phases, several of the headlining speakers weighed in on how we might reconceptualize the problem of how life arises from nonlife. Most of the really engaging discussion and debate centered on the topic of screwdrivers. Yes, screwdrivers. You'd be shocked how many times these handy tools are invoked in deep conversations about the nature of reality—almost as many times as they are used to remove bathroom doorknobs when toddlers lock their parents out.

I think the reason for this specific example's popularity is traceable to one charismatic source—the mind of Stuart (Stu) Kauffman, who was in turn probably inspired by the philosopher Ludwig Wittgenstein. Stu has a fascination with screwdrivers as a representative object *with function* that exists within in our biosphere. He argues that all the things that the biosphere has created could not, even in principle, be predicted a priori, because they have a function.[19] Stu likes to talk about it in terms of affordances—the possible actions of a given object. You cannot predict all the affordances an object like a screwdriver might have, because you do not know all the objects (whether they currently exist or not) that a screwdriver might interact with. Indeed, in Stu's view the set of possible objects the screwdriver could interact with is infinite, and therefore not prestateable.[20] It follows, then, that you cannot predict the future evolution of the biosphere because there is no way to iterate over the entire space of what might happen. Stu uses this to argue that life cannot be described in a lawlike way.

I greatly admire Stu and his insights, but on this point I disagree. We do not need to iterate over all the functions a screwdriver or any other object might have, because the vast majority of those are irrelevant to the future of our biosphere. Most of the objects that Stu imagines as not prestateable have no possibility to ever exist, so we should not count them. The function of a screwdriver is not a property of the screwdriver—it is a local, relational property of all the objects the screwdriver comes into existence alongside in the unfolding of our biosphere. Thus, if you study the objects that exist now, they should already encode all the information you need about the functions that constructed them to also anticipate much of how the near-term future will be constructed (though not all if our universe can produce genuine novelty).

Function is notoriously tricky to define, but a good working definition is in terms of the relationships between objects that arise due to selection. Some things can exist only if something else already exists that makes their existence possible. That is, possible physical objects need other objects to exist to themselves be selected to exist. Since any object that exists is itself finite, the number of things that can come into existence at any time must also be finite, and therefore, at least in principle, stateable.

Now, here is a question for you: If you found a screwdriver on Mars would it be evidence for life? If so, why? If not, why not?

Many of us are perhaps not yet willing to ascribe the term "life" to a screwdriver. But cast in the question of whether it is evidence for life, we are probably willing to admit that yes, yes, it is. This is because we do not expect screwdrivers to spontaneously assemble, or fluctuate, into existence based solely on what the laws of physics and chemistry can accomplish, starting from the geochemistry of Mars. An evolutionary chain of objects is necessary to assemble screwdrivers into existence.

Complex objects do not come into existence without the information first existing to make them. The question then is where this information comes from—is it designed or evolved, or both (e.g., by evolution generating things that are capable of design—things like us)? In current physics, no "information" arises in the evolution of the universe: all information specifying why some things exist and not others must be encoded in the initial state. The only selection that can happen is in the choice of the initial state and the law of physics: this defines how everything behaves forever. Explaining the initial state of the universe is problematic in the fields of fundamental physics and cosmology, and here we see plainly why: we have no mechanism for why a choice was made for one initial state over another.

Charles Darwin poignantly captured the stark juxtaposition between the physicists' view of reality and what is necessary to explain life in *The Origin of Species*:

> Whilst this planet has gone cycling on according to the fixed law of gravity, from so simple a beginning endless forms most beautiful and most wonderful have been and are being evolved.[21]

When I say selection, I mean it in the way Charles did, but also in a more generalized sense—it is not just biological species that are selected, but everything. Selection and existing are the same thing. For complex objects—like biological and technological ones—existing is very rare in the space of possibilities when we consider all the things that could exist but don't. So selection is both more apparent and necessary as an explanation for the existence of objects the more complex they are.

Charles's is such a beautiful turn of phrase I had for a long time assumed he was a universally beautiful writer. I recently sat down and attempted to read some of his other original works in earnest and realized how sorely mistaken I had been! They're also brilliant, but often a dense read. Even so, Charles shows a stroke of genius here both in prose and in thought. He is making a fundamental observation of what Isaac Newton's physics (and by extension modern physics) is missing in its ability to explain life. In a physics with life, we should expect an endless succession of novel forms, each constructing the next. Screwdrivers emerge along with screws and a slew of other tools on planets where life has also emerged and evolved intelligence; they don't fluctuate into existence on their own as an individual instance and at random, as some theories of modern physics suggest they could (albeit with low probability).

But in the physicist's current model of the universe, everything runs by the ticking of clockwork mechanisms that can trace the origins of all creativity to random fluctuation, or to the initial state of the universe.

Resolving the dichotomy between the Newtonian and Darwinian descriptions of the world is key to understanding the deeper physics governing the origin of life. This is because the origin of life is the unification point between biology and physics. It is where the universe described by the fixed laws of known physics—a universe without us—must yield to the seemingly endless forms of complexity generated in the evolution of our biosphere or any other. This unification must happen in what we call chemistry, because chemistry is the first thing the universe builds where not every object can exist (recall how large even small molecule space is). This leads to the possibility of an unfolding of different forms in different locations—what we might call different instances of "life." Not all chemical possibilities can exist all at once; the ones that do exist *must* therefore be selected. This is why chemistry also happens to be where life can first emerge.

The challenge with unifying different conceptual paradigms is that it often requires building a new one. You might envision our knowledge as a mosaic, displaying a picture of our current understanding of the world, but there is nuance: if you look at it in the right way you can see how you could take all the current pieces and build a completely different picture that is just as consistent. So how do the pieces we have now fit together to build a new mosaic?

Focusing on the problem of what can exist, and when, resolves much of the ambiguity about the causal power of information—that is, that information can cause things to happen. Saying information is causal merely points to the fact that some things that come to exist require something that already exists to assemble them. Some require longer chains of causation and are thus more "informational." When we discussed the hard problem of matter, we recognized that it is impossible to know what something is intrinsically, in the absence of interaction. We recognized that what we can describe as existing—from the perspective of what we can interact with and measure—is the software of reality, the interactions that make objects or allow us to observe their existence.

This opens a possible resolution where we must treat *matter itself as information*, which forces reconsidering our concept of what is material. In some ways we are reenvisioning what the quantum physicist John Wheeler envisaged with his famous dictum, "It from bit," where his "bit" refers to the measurement of information. However, what I am claiming here is that if we adopt John's "it from bit," the most interesting feature is not that what we call reality arises from the posing of yes/no questions and the registration of measurement-evoked information theoretic responses to them, as John claimed (where 1 bit is required to measure the answer to a yes/no question). Instead, what I am suggesting is that what we call the future is constructed by those responses. This property must not be merely a choice but a material property of the objects we observe. This brings us to what we discussed with respect to the hard problem of consciousness, where we flipped the question of what it is to experience existence around, when we asked if there are any measurable dynamical consequences if we assume physical systems exist that do have inner worlds. From this perspective, the key feature is the ability to imagine or represent things that do not exist, but could, such that the act of imagination becomes causal to the existence of some objects. In the language of life and information I just introduced, we would say that what we call consciousness must be a feature of the causal chains necessary for the formation of some objects. From this extrinsic perspective, consciousness is a particularly concentrated example of what life is doing more generally.

If we are ever to understand what life really is, we need to recognize that among the unimaginably large number of things that could exist, or even the smaller subset of ones that we can imagine, only an infinitesimal fraction ever will. Things come into existence when and where it is possible to—and what we call life is the mechanism for making specific things possible when the possibility space is too large for the universe to ever explore all of it.

INTERLUDE

NEW PHYSICS

Your lineage takes pencil to paper. Scribbling notes, you—as you are in 1910—are fiercely trying to elucidate how it is that gravity can work in situations where the speed of light is constant. You are frustrated and have many conversations with your collaborator, Marcel Grossmann, a brilliant mathematician and friend. Still, it takes you nearly a decade to formalize your intuition. The idea was there all along. However, translating it so others among you could readily see it—and more importantly *test it*—was far more challenging than having the intuition itself. Even so, you persevered and did eventually succeed in formalizing your thoughts. Your theory was validated three years later, after the intervention of a global war. That experiment was led by Sir Arthur Eddington, who had the stroke of insight to use the solar eclipse to test his predictions. The eclipse obscured the light of our Sun and allowed direct observation of the bending of the light from distant stars around it, just as your theory had predicted we should observe it.

It would take decades after your death for us to later prove other predictions your theory made. A key implication of your intuition was the existence of gravitational waves—ripples in the fabric of the space-time you invented to describe how our reality works. The first compelling evidence that your gravitational waves might really be there came in 1978 when two humans you would never know, but who knew of you, named Russell Hulse and Joseph Taylor, observed a binary neutron star system and inferred that the missing energy must be dissipated, exactly as you predicted, by gravitational waves. Like you, they were awarded the Nobel Prize; theirs was for confirming some of your predictions. However, their evidence was not direct. Our quest continued to validate what you intuited was there. In the 1990s, roughly forty years after you left us, we started to build interferometers to directly test your predictions. These devices needed to be so precise it took us decades to build them, even with our technology as it exists now, which is exponentially more advanced than anything you had ever witnessed in your lifetime. But we, too, ultimately persevered. On September 14, 2015, we made first contact with the phenomenon your intuition told us existed.[1] We do not have to tell you this is one hundred years after you formalized your ideas! Did you and Marcel think it would take so long? Perhaps you would think this was quick.

The measurement was among the most precise made in our history and indicated the presence of two colliding black holes with masses roughly thirty and thirty-five times our own Sun's mass. The ring-down of this merger on our detectors, the symphony of a universe describing itself, lasted 0.2 seconds and was sung on our instruments roughly 1.4 billion years after it happened. Albert Einstein was not born yet, and neither were we: the lineage of life we share was not even multicellular then. His theory was such that it predicted features of events that happened in his past, which could be confirmed by us only in his future.

Our universe is such that this is possible. By observing what is happening now, in the present moment, we can infer what has happened already, even billions of years in the past. We can predict and even cause what happens next. We can write down theories that describe how our reality works, and then test them against observations and experiments. Theoretical physics, among human cndcavors, has bccn stunningly succcssful at this, in no small part because of minds asking the hard questions about the nature of reality. These are minds like that of Albert Einstein, who painted our current understanding of the nature of space-time with his general theory of relativity that led to the prediction of gravitational waves.

Albert was once a part of the lineage that is human, with all that entails. He enjoyed playing violin, he was fiercely curious and intrepid in his pursuit of understanding reality, he believed education should be accessible to all, and he advocated for an end to systemic racism in the United States—often visiting historically black universities. While living, he played a pivotal role in our current understanding of gravity, space, and time, and even laid some of the foundations for quantum mechanics. He also made significant contributions to philosophy, morality, and ethics. Like every human, he also had his flaws—historians have written about how he was a mostly absent father to his two sons and had multiple affairs over the course of his two marriages. All these details of Albert's life are fiercely important, as are the details of any life.

They, however, have less relevance to the fundamental physics of what it is to be alive, because we are not looking at what is specific to us as individuals, but what is universal to us all. What part of you is also a part of Albert Einstein? What part is also a feature of the very first life on Earth? By all I really mean *all*—not just the instances that we exist as today, but what we have been over the last several billion years and what we will become. We are each just one temporary instance of life on this planet—a pattern of information structuring matter across billions of years.

To get to a universal understanding of what life is, we must do what theoretical physics does best—abstract the nature of reality to its most fundamental and powerful explanations. Universal includes us humans, our ancestors—from microbes to multicellular life—artificial life, artificial intelligence, what comes after those, and even aliens.

The most important thing about Albert Einstein for our purposes is that he was not only human, and a theoretical physicist to boot, but also an example of a broader set of phenomena that exists in our universe that we have yet to understand—like you and I, *he was once alive*. The theories he produced, that tell us so much about the nature of reality, would not exist if there was not a part of the universe—life—that can comprehend the rest.

What we are about to embark on next is a vision of a future where we understand what we are, not based on theoretical physics as it was—the kind of physics Albert Einstein devised, with his predecessors Isaac Newton and James Clerk Maxwell, his contemporaries Marie Curie and Emmy Noether, and his successors Richard Feynman, Steven Weinberg, and Frank Wilczek, among many others. They all studied the nonliving universe: the universe without us.

Instead, we are now stepping into a possible history of what physics will be if we are to explain the most intimate of physical phenomena—the nature of ourselves. You are a part of that lineage, too, and it is only by accruing knowledge over our lineages that we can understand how it is that we came to be.

Three

LIFE IS WHAT?

There's an old joke about a physicist talking with a neuroscientist. The physicist would like to learn more about the brain. The physicist says, "Tell me something about the brain." The neuroscientist says, "It has two hemispheres." The physicist says, "Stop! You've told me too much!"

I am sure my biologist friends will be demoralized by the details I am glossing over in this book, and indeed the many details that form the richness of life on Earth that I cannot touch upon. I am at times demoralized too.

But the true art of theoretical physics is to unveil what we cannot directly observe with our senses, by cutting through the extraneous details to understand the patterns revealing the essence of things. Our project here is not to attempt to explain *all* the detailed and beautiful variety of life—that is an impossible task for a theory builder. What we are doing is attempting to explain what all life in the universe, even that which might lie beyond

Earth or exist only in the past or the future of our biosphere, should share. To be so broadly explanatory, these features will necessarily be very abstract.

Right now, as I write, I am sitting in my backyard affixed to my chair because of gravity. I look up and see the Moon, fixed in its own orbit by gravity too. Without knowing about gravity, it would seem truly fanciful to imagine the reason I stay in my chair is the same reason the Moon stays in orbit around the Earth. To first figure it out, Isaac Newton had to ignore almost all details of the objects in question. It does not matter what I wear, how I sound, what I feel (even though those details may be important to me). It does not matter that the Moon is primarily composed of oxygen, silicon, magnesium, iron, and calcium, or that it—presumably—has no feelings. It does not matter that the Earth has a core made of iron, that 71 percent of its surface is covered by water, that a species we call *Homo sapiens* first emerged on its surface around three hundred thousand years ago. All that matters to calculate the gravitational force between me and the Earth, the Moon and the Earth, or me and the Moon is each of our masses and the distances between our centers of mass. This is the content of the universal law of gravitation. It played a central role in the unification of motion in the heavens and on Earth; it's all governed by the same rules. While this unification is obvious to us now, for hundreds of thousands of years this knowledge was hidden from our ancestors.

This is the great trick physicists use to get to the heart of reality: we identify what details are important and drop all the rest. This art allows us to unify a seemingly disparate range of things in a mathematically concrete and testable way. When talking about motion, we do not concern ourselves with the details of different moving things. To calculate motion or gravitational attraction, we need only concern ourselves with mass, position, and acceleration. This simplification was a huge conceptual leap made by our species—it allowed us to describe all motion, whether here on Earth or on the other side of the observable universe, in the same way.

It's not that the idiosyncratic details of biology as it evolved on Earth don't matter. They just do not matter if you want to understand life as a universal phenomenon. To do that we need a description abstract enough to unify all the things we think are life, anticipate the things we don't think are life but are, and solve the question of life's origin. What I mean by this is that it must be a sufficiently deep conceptual structure to unify what happens in physics and chemistry with what happens in biology. At the same time it must allow us to recognize alien examples of life, which means that the abstraction has to apply predictively to all life in the universe, not just life on Earth.

What details are important? Which ones can we neglect? These are challenging questions.

What is life? Well, that's a really challenging question. The answer we eventually arrive at need not conform to the expectations we have now. In fact, historical precedent suggests that the best explanations in science often do not conform to our initial, naïve expectations. Rather, we should ask "Life is *what*??" preemptively anticipating our own surprise in the answers we will find.

In the rest of this book, I'm going to lay out one leading set of ideas on how we can start thinking about building a theory of physics to explain life and its origin. This approach is intended to capture life's universal yet very abstract properties in such a way that we can design new experiments to test our ideas. We are operating in the same spirit that led to paradigm shifts in our understanding of the physical world in the past. We should expect to be surprised. Indeed, my goal in writing this book has been to build a surprising vision for what solving the origin of life, and by extension what first contact with alien life, might look like. We need to imagine what is possible before we can cause it to exist.

A Theory of Life's Objects

I've already hinted at where we are going: the new physics needed to explain life starts with understanding what exists and why. That is, it starts with selection of, and for, what gets to exist as physical objects. Existing is special: most things—at least among the set of the ones that we can even imagine—will never even have the chance to exist. Perhaps even more things will never even have the chance to be imagined.

But let's start with the things we can imagine, because some of those objects do at least have some hope of existing. We can imagine, for example, a set of elementary particles, then stepwise we can construct a space of their combinations to assemble atoms, molecules, cells, bikes, people, plants—all the things we have encountered in our planet's four-billion-year history, and all the things we did not encounter but could have. This is a small part of the space of what is possible, and it is the space we live in.

Modern theoretical physics has three primary pillars—quantum mechanics, general relativity, and statistical physics. Quantum mechanics is concerned with explaining the properties of matter and light. General relativity and special relativity explain gravity and motion and do so by defining coordinate space and coordinate time. Coordinate space is the technical term physicists use to describe what we call physical space, defined by the three dimensions you can move in, i.e., left-right, forward-backward, up-down. Coordinate space is what rulers measure. Coordinate time is the technical term physicists use to describe what clocks measure: the regularity of clocks is expected to record the passage of time more reliably than our perception of it. Statistical physics and thermodynamics describe the properties of energy, work, and measurement uncertainty (entropy).

We have not yet built a physics that deals with the combinatorial space of all possible objects that can be built from elementary stuff. An example is the space of all possible molecules that can be assembled from the periodic table of the elements. You might imagine that in a different universe there could be a different periodic table or none at all, indicating the chemistry we have is contingent on the properties of atoms here in this universe. Another example is all possible technologies that can be assembled with transistors. If another technology had come along before or instead of transistors, the set of technologies that exist now would be entirely different. The spaces of all combinatorically built objects are far too large for the universe to make everything in them, or even a small fraction, at any given time. Most of the things that could be made will never be made. This is more true for complex things made out of many parts, because the more parts there are to make an object, the larger the possibility space that object exists in.

Those possible things that get the privileged status of becoming physical objects are contingent on what came before, and it is precisely the chains of contingency necessary to assemble them that cause these objects to be far too improbable to be explained within current theories of physics. I'm going to make the argument that in order to explain these objects, we have to reconceptualize the space of combinatorial possibilities as a physical space—and it must be just as physical as the three dimensions you can move in. Understanding the physics of this space is key to unlocking the mystery of life.

The screwdriver is one possible object in this space that we know of that can exist as a physical object in our universe. We know this because we have observed it does exist. We already asked: Is it improbable or impossible for a screwdriver to spontaneously appear on the surface of Mars?

What about *two* screwdrivers?

The meeting at the Carnegie Institution I mentioned in chapter 2 may have planted the seeds for a paradigm shift in how we approach the origin of life. It is already allowing some of us to see life, and the universe, and everything, in an entirely new way. We are doing so by uncovering new bridges between the realms of physics and biology. As with most unifications, the new paradigm looks unlike the former ones it emerged from.

Among the debaters at the Carnegie meeting, one individual boldly claimed to be able to prove that observing a screwdriver forming by random fluctuation—that is, without an evolutionary process—*is physically impossible.* Furthermore, he claimed this could be proven, not logically, but experimentally. That provocative statement was made by the University of Glasgow chemist and polymath Lee Cronin. Lee is a friend and collaborator, and he is often the most vocal person in the room trying to push boundaries of what we know or assume.

Lee could be considered a theorist, except that his method of theory building is extremely unusual: the theory must be something he can manipulate in the lab. While most of us build theories by manipulating abstractions in our minds, Lee builds theories by manipulating physical objects in the real world. What he meant at the Carnegie meeting was not that he could prove the impossibility of a spontaneous screwdriver per se (as that would be hard indeed!), but that he could prove the idea experimentally, for molecules.

Lee will often join Zoom calls from his home workshop, where you can see experiments running in thc background. On our regular Friday calls during the pandemic with two other friends and collaborators, Chris Kempes and Michael Lachmann (both at Santa Fe Institute), Lee would show us his latest home experiments. Lee's home workshop includes 3-D printers, robotics, electronics, and I am not even sure what else; he's constantly acquiring new tools and devices. His experiments are often aimed at troubleshooting ideas before he brings them to the lab. One of the ones he was developing during the pandemic was an at-home selection engine using a 3-D-printed arena where patterns assembled from ball bearings would compete and be selected among. The aim was to test our conjecture that evolution and selection represent universal physics. If we do figure out the basic principles of the laws of physics that govern selection, it should be possible to show whether the physics is really universal even in a simple, tabletop experimental setup. Lee also attempted to make his own quantum computer. He later joked about how he may have achieved an entangled state for the photons in his device but was not able to prove it—the device that would allow him to perform the quantum measurement would have cost in the ballpark of one hundred thousand dollars (a little steep for a backyard workshop). He subsequently published related experiments with collaborators in the physics department at the University of Glasgow, where they used a polarization-entangled photon source to measure optical activity with unpolarized light.[1] Lee often has movies related to the detection of alien life, such as *Arrival* or *Contact*, projected on the back wall while he works. Sometimes he has two different movies playing simultaneously on different walls, which I know I'd find impossibly distracting if I was attempting to focus—it's distracting enough for me when he has one movie playing during our discussions!

In thinking about how to build a way to measure life, Lee made a conjecture: like screwdrivers, cell phones, satellites, and cars, some molecules are complex, improbable objects too. That is, he recognized that chemistry is the first place where the universe cannot generate everything. Some molecules might be so complex, they are surefire signatures of a process that had to evolve to make them. Such objects simply have too many parts to have formed spontaneously.

If we could build a theory based on Lee's intuition in chemistry, we might be able to formalize the foundations of a theory for life not in terms of what life is, but of what it does. In chapter 2, we discussed how shifting focus from what life or consciousness is, to what these do, might open new ways to solve hard problems. The idea is pretty simple: *Life is the only thing in the universe that can make objects that are composed of many unique, recursively constructed parts.*

Solving the origin of life by testing a measurable theory in the lab is not a straightforward thing to try to do. Many scientists think it will be impossible to understand what life is, let alone develop a deep theory of the physics underlying life that we could test in the laboratory. Some think we are crazy for trying, or that the ideas are incomprehensible. Even worse, some think them trivial. I can understand how our ideas can seem this way to some, because they exist in a creative space that is totally alien. Since we have not solved the origin of life yet, nor have we ever made contact with aliens, maybe what we need are some radically new—and therefore alien—ideas.

I was motivated as a PhD student to study the problem of life because it was uncharted territory. It was clear we did not know what questions to ask, let alone how to answer them. I admired the huge conceptual leaps in our understanding of physics in the past, including ones I've already mentioned in this book, like how Isaac Newton discovered gravity, James Clerk Maxwell unified electricity and magnetism, Albert Einstein unified space and time, and how quantum physicists had revolutionized our understanding of certainty and locality. Each of these developments, and every other major shift in physics, has come from looking at an unexplained phenomenon with fresh eyes. I started working on the origin of life because I thought there might be hope we could do this for life too. But as a classically trained theoretical physicist I did not think the ideas would be testable in my lifetime. Lee aimed to prove me wrong. Early in our collaboration he pushed me to ground my thinking in what we can measure, and this exercise has made all the difference in the reality I now see written in the story of life.

With our work on assembly theory, we are not attempting to define life—in this book, you and I have already seen how such attempts fail. Instead, our aim is to provide a formalism for unifying life and the inanimate. How can we uncover the physics that gives rise to life and allows the transition from nonlife to life to even happen?

The key hypothesis of assembly theory is that the observation of complex objects implies that the "information" about the steps of their formation must exist in other objects—that is, it implies a physically instantiated memory—or, if you like, a set of constraints—for their formation, which can happen only via evolution or learning. We talk about this more generally in terms of selection as a unifying conceptual paradigm: if selection among the space of what is possible happens before life, the constraints on the space of possibilities can drive the emergence of what we observe in life as evolution. Both Michael and Lee will talk about this as "selection before life" and say we need to discover the laws of physics that explain selection. I like to say we are looking for the physics of what gets to exist.

To get to a testable theory, we are doing something unusual: we are working backward. We're asking what we can measure about molecules in the lab that would indicate when they are objects that can emerge only via selection and evolution.

Assembly theory is a theory of objects, because objects are what we can measure. In my entire physics education, it was never clearly stated what an object is. Instead, I implicitly picked up an understanding that the only fundamental objects of interest should be those that are indivisible: what the Greeks called atoms and what we now call elementary particles. In assembly theory, our concept of a fundamental object is entirely different. Here is how we conceptualize them:

> **Objects are finite and distinguishable.** A key feature of our universe is that we don't see a smear of everything existing all at once. There are a finite and countable number of objects, not an infinity of them. The things we observe exist as discrete, distinguishable objects—we observe basketballs and baseballs but not all possible objects that could exist in between. This is true even at the level of elementary particles.*
>
> **Objects are breakable.** In the history of science, what counts as an indivisible, and therefore fundamental, object has always been defined by technological and epistemological limitations. But this historical contingency is seldom, if ever, explicitly acknowledged. An example is how when atoms were first discovered we thought they were the smallest building blocks of all things. Later we discovered atoms are made of neutrons, protons, and electrons, and subsequently that even the neutrons and protons have substructure—they are made of quarks. If string theory is successful, it would replace the current set of elementary particles with a new underlying mathematical structure. Confirming this requires technological advances beyond our current capabilities—a clear example of how what we con-

---

*In the case of quantum physics there are such things as indistinguishable particles. For example, particles known as bosons—things like photons—can occupy the same quantum state at the same time and therefore are indistinguishable. However, this is a collective state of these particles and should be considered as only one object from the perspective of assembly physics.

sider fundamental keeps pace with technological advance. What is considered fundamental in modern physics merely defines the boundary of what we can observe.

In assembly theory, we do not define objects this way (as what we cannot break down further). Instead we define objects as all things that can be built from elementary building blocks using operations that are consistent with the physics of our universe. Objects are therefore the opposite of what they are in physics: they are the things you can build and break apart. This allows everything made of elementary parts to be objects with causal power in assembly theory. It is one reason assembly theory naturally accounts for "emergent" properties: it defines as fundamental all things that traditional physics would say are emergent. It opens the door for features like agency and free will to be explainable when Brian Greene and others argue against them, because current physics puts all causation at the microscale. In fact, in assembly theory we regard almost no causation to exist in elementary objects, because causation is built up over time along lineages. I regard this as a necessary feature of life: the objects it describes must be able to evolve and change as the life that constructed the theory itself evolves and changes with its advancing technology.

Objects exist more than once. The idea of copy number—how many of a given object you observe—is of fundamental importance in defining a theory that accounts for selection. This is because the more parts a given object has, the less likely it is that an exact copy of that object can exist without some precise mechanism that has itself been selected and that generates the object. Assuming even a single object could potentially be generated in a series of random events, the likelihood of producing one copy of the object becomes exponentially less likely as the number of steps to assemble the object increases.

This is doubly so for two copies of the object, and so on. A single example of a highly complex object does not provide evidence of evolution or selection. In assembly theory we push this further—we assume objects do not matter to measurements unless they exist in more than one copy. This is certainly true for molecules, because to measure the properties of a molecule in a mass spectrometer (the instrument we use to measure assembly), you must have many thousands of copies of the same molecule.

You may regard this as a leap too far: Surely an individual human is measurable and need not exist in two copies to be an object? But this is not what the theory is saying. We already know that any two humans sampled at random will share 99.6 percent of their genome—our variation genomically comes from only 0.4 percent of our DNA. Likewise, we are similar in more ways than we are different in the vast space of possibilities in every property we might observe ourselves: our shared culture, environment, ideas, languages, fashion sense, etc. In assembly theory we recognize that all humans are nearly an identical object because of the history we share—the present moment is only a tiny piece of the billion years of acquired memory that built us.*

Our emphasis on copy number is the most intuitively difficult feature of assembly theory for most people, and it is perhaps the most unique feature of the theory. Beyond measurement and requiring evidence of selection, there is, in fact, a deeper

*Saying "all humans are nearly identical" does not mean that we do not display a huge amount of variation: what it means is that every difference you have ever observed between yourself and another human being defines a volume of the space of what is possible that can feel large to us from inside it, but if we compare it to just the size of the subset of possibilities we can even think about, we find it is very small indeed.

reason we think objects must exist in more than one copy: if the mechanism has been selected to make them, they can be made again. Objects existing in multiple copies have become a regular feature that is reliably reproducible in our universe, in a manner like what we discussed in chapter 2 for the regularities associated with launching artificial satellites.

Objects are lineages. For the lineage connecting all life on Earth that goes back nearly four billion years, humans emerged as an object with features distinct from the rest only between five and seven million years ago: we are distinguishable from the rest of life on this planet only in the most recent 0.2 percent of life's history on Earth. We already noted how the steps along the causal chain leading to each of us is nearly identical: we are the same lineage. In assembly theory, objects are their lineages: these lineages, or causal chains of objects making other objects, are folded up today and persist in objects that exist now. We formalize the lineage as the history of steps necessary to make humans possible: an instance of that lineage is any individual one of us.

Objects form via selection. The final property we assume of objects is that their mechanism of coming into existence is selection. An object can take on the other properties I have just specified only if the universe has sufficiently constrained the space of possibilities to make that specific object at a particular point in space and time. This also imposes locality for objects—objects can appear with other objects only if they share a common lineage in the past: that is, objects can exist in proximity to one another in space and time only if the causal chains leading to their formation share some history; that is, they must also share a locality in how they are constructed within the space of all possible things that could exist.

The definition of objects I have just run through is very abstract. But this is essential. In describing objects abstractly in this manner we are also being concrete about the features we expect to be important in building a theory about why some objects exist and others do not—i.e., a fundamental theory of selection. We are now ready to talk about a theory of physics that considers life's objects as fundamental.

Assembly Theory

Recall our conjecture that *life is the only thing in the universe that can make objects with many unique parts*, and also that we might use this as a means to look for evidence of selection and evolution in measurable properties of molecules. I said that objects must be breakable, and therefore that they are also buildable. The *assembly index* is how we formalize how hard it is for the universe to build something. We calculate it by breaking an object apart: it is defined by the smallest number of physically possible steps necessary to produce an object. For a given object, the *assembly space* is defined as the pathway by which that object can be built from elementary building blocks, using only recursive operations combining what has already been built in the past. In the case of the shortest path, the assembly space captures the minimal memory, in terms of the minimal number of operations necessary to assemble the observed object, based on objects that could have existed in its past. Importantly, assembly spaces are constrained by the laws of physics: possible steps in an assembly pathway are ones that must be physically possible.*

Imagine you have a hypothetical molecule, and you divide it into two parts by breaking bonds in the molecule. You do this again and again until you get to very simple building blocks—in the case of molecular assembly theory, these simplest objects will be made of atoms and bonds. This is because possible molecules must obey the bonding rules of chemistry: if you want to look for evidence in a molecule that it is a product of selection, you must use the kinds of operations a constructor would constrain to assemble a molecule. Here a constructor is any physical system with information on the constraints necessary to localize the space of possibilities to be specific to formation of the object in question: for chemical objects, the constructor could be a geochemical environment, a chemist, or an AI-driven robot. This is consistent with the notion of "constructor" that I introduced in chapter 2 when discussing Chiara Marletto and David Deutsch's work on constructor theory. No matter what it is that builds the object, it must do so using operations that are possible. To start to build the assembly space for our molecule, you can take your set of simple building blocks, and by joining a pair together (e.g., making a new bond), attempt to reassemble the original molecule. The rule is you can select any two molecular fragments to combine at each

*Here I mean physically possible in the sense often described by David Deutsch, also a key element of constructor theory he is developing with Chiara Marletto. There they do not restrict themselves to "physically possible" pertaining only to laws of physics as we understand them now, but the more general concept of whether a given operation is possible in our universe.

step, as long as you made those two fragments in the past. The joining of parts is a recursive operation, allowing reuse of previously built parts at each assembly step. The assembly index is the length of the shortest assembly pathway within the molecule's assembly space.[2]

Molecules are not intuitive to most of us, so let's try this again, this time with LEGO. We will lose little of the relevant physics. If you want an excuse to play with LEGO, pull some out and try this yourself. Imagine (or build) a simple object, say one that has two blue pieces and three yellow—whatever shape you like. This is your object. Now, take it apart into individual building blocks—the nondivisible LEGO bricks are your "atoms" and they come preconfigured with "bonds" because of the structure of how they can attach to one another. Let's assume you have access to unlimited LEGO pieces—right now we're defining the abstract space that is a LEGO object in assembly theory, not in terms of the actual resources (number of blocks) necessary to bring such an object into existence by building it. You can think of the assembly pathways as the set of all the ways the object could be caused to come into existence. To build an assembly space for your LEGO object, you can start by taking two structures from your pile. With your target shape in mind, put two pieces together that would fit into your original design as a subpart. Add the structure you just made back to your pile. Next, draw two new pieces from the pile, where one or both can be the piece you just built.

You can keep doing this recursive operation—reusing any pieces you have already built—until you have assembled your original object. The recursion is important because it implies you can only use objects you have already learned how to build—that is, assembly spaces carry with them implicitly a notion of memory of the past. You cannot use an object to make something new until that object itself exists. Seems logical enough, right? Yet no current theory of physics contains this kind of physical embodiment of past histories in the present. In assembly theory, the past is rolled up into the present, and is instantiated in the current moment in the assembly space of objects that exist now. Stated slightly differently, assembly theory describes every object you interact with as a recursive stack of its history. You might think of assembly theory as a theory of physics where history is a physical attribute.

The set of steps you would have taken (or maybe did take, if you made your own LEGO version) is what we call an assembly pathway. If you ran the process again, you could follow another pathway. The assembly space we care about in assembly theory is the one with the shortest number of steps: this assembly space minimally captures the amount of causation necessary to make the object. Determining the assembly index is straightforward when you have the minimal assembly space: it's just the number of steps on the path. That is, it is the minimal number of operations you could take to make your object by reusing only unique shapes you have already made. You can think of this as the minimal history, memory, or (causal) time necessary for the universe to assemble your object. Again, an important feature is that the universe cannot invent anything new until it has produced the parts necessary to make the new thing. There is no free lunch that spontaneously fluctuates into existence. Only evolution and selection can produce your lunch.

Now, because LEGO can attach only in certain ways, you'll quickly notice you are restricted in the shapes you can build. It is hard to build a circle from square blocks, for example. Anyone who has ever played the popular video game *Minecraft* will know this well. In a universe made of blocks it is much easier to make a square Sun than a round one.

There are two constraints we recognize in the physics of assembly theory: (1) the laws of physics that govern the assembly process (e.g., the laws that determine what can happen in our universe), and (2) the information that is the constructor of the object (e.g., you in the LEGO example), which must be remembered for the object to be generated. The LEGO object cannot exist without something like you to build it.

For the first constraint, it is important to acknowledge that we cannot arbitrarily stick blocks together willy-nilly. Well, maybe we could if we had superglue, but that would violate the rules of the LEGO universe in our example above. The constraints on how LEGO blocks can stick together are a bit like the bonding rules of chemistry, which apply as universal rules in our real universe. Not everything is allowed, and this restricts the set of things that can be assembled—they must be possible to build based on the laws of chemistry (e.g., what bonds can form between atomic elements).

The set of possible objects you can construct is also limited by your imagination, or more specifically by the memory you have of LEGO-like objects from which you were able to combinatorially imagine new possible objects. The object you constructed was already a member of a restricted set, with constraints imposed by your memories (which bound the space of your imagination), as well as what you have time and resources to build.

Objects with a larger assembly index take more minimal steps to build, and there are also exponentially more of them. We are playing with just a few blocks but imagine if I had asked you to build a LEGO Hogwarts as your target object, by following every possible path to construct it. There would be a lot of paths indeed! You would still be going. The shortest path would be quite long (I have not personally calculated the assembly index of LEGO Hogwarts), and whether you got there at all would depend on how much of a shared history you and I have, reflected by whether you even know what I mean when I say "LEGO Hogwarts."

The shortest path is not necessarily the one the universe takes to make your object. In fact, the number of possible pathways grows exponentially with the assembly index of an object. On probabilistic arguments alone, it is very unlikely a minimal path would be the one by which any complex object is discovered. Additionally, in chemistry, it is worth pointing out that assembly steps are not the same as reaction steps. Some objects within an assembly space are only fragments of molecules and could never exist as isolated objects, so would never appear as a finite, distinguishable object in a reaction sequence. Yet we argue they are critical to understanding how much causation is necessary to making the molecule, because they are an element in the memory of its possible construction, and this is independent of the particular reaction sequence that makes the molecule. You may think it odd to have a theory that includes objects whose properties can never be measured in isolation, but in fact this is fairly standard practice when it yields the correct empirical results and has high explanatory power. Our ability to build explanations on such abstractions is one of the most important tools our species has invented for inferring the presence of physical things that we cannot directly observe in isolation. A good example are the quarks, elementary particles that play a prominent role in the standard model of particle physics. Quarks are what other important particles—like the proton and neutron in atomic nuclei—are made of. But by the very physics that governs quarks (as we currently understand them), they can never exist as isolated objects and must always exist as a grouped property of the objects they make up (this property is called confinement). This is true for many of the fragments of molecules in molecular assembly: these fragments are critically important to identifying the amount of selection, or causation, in a configuration of molecular objects, but may never exist outside those objects.

To give an explicit example, the assembly index of the molecule adenosine triphosphate (ATP) is 21. ATP is a key energy carrier in the cell and can be made by several different metabolic pathways; the most well-studied include glycolysis and the citric acid cycle. These reaction pathways are not the same as what we talk about in assembly theory. You can think of a reaction as a coarse-graining, or systematic grouping of the transformations that can happen in the assembly space, that depends on the particular constraints that cause the reaction to occur (e.g., on the constructor). These are different in different parts of the universe. For example, these constraints might be a rock surface in geochemistry (well, not for ATP, but other simpler molecules) and a round-bottom flask in a chemistry lab, which are both different from the orchestra of enzymes that mediate the reactions performed within a cell. Because which substrates and reactions are used to make ATP is context dependent, the reaction sequence is not an intrinsic feature of ATP. You might be able to make ATP in a reaction sequence that takes a couple of steps, but in those cases, you are always starting with other molecules that themselves have a relatively high assembly index: such molecules already represent a huge set of constrained possibilities, which have become embodied in a selected object.

We assume the assembly space is intrinsic; it tells us the same things about the molecule anywhere in the universe, independent of whether a microbe from Earth or an alien made it. The assembly index is therefore universal—if you found ATP at the bottom of an ocean on Earth or in the ocean of Saturn's moon Enceladus it would have the same assembly index.

You probably did not follow a minimal path the first time you assembled your object (although our initial LEGO objects were very simple, so maybe you did). In assembly theory, we care about the shortest path, not the path the universe did take or the average of the paths it could have taken, for a few important reasons.

The first is the point I just made—we regard the minimal path as an *intrinsic* property of the object. If we picked two of you reading this book who by chance built the same object, you probably took different pathways. The specific pathway you used to build your object is context dependent—it depends on both the constructor (you) and the object. I am emphasizing this point because one of the biggest challenges in defining "information" in a biologically relevant manner has been how "biological information" appears context dependent: the bits of "information" in some sequences of DNA (e.g., genes) "mean" something within the context of the cell (they provide a function), and for other sequences there is no function. Furthermore, the biologically meaningful sequences do not have a function outside the context of the cell. You could not tell the difference between a DNA strand with function and one without by just looking at the sequences of A, G, C, and T; this has been one of the most challenging aspects of defining a biologically relevant concept of information. What we do in assembly theory is different. We remove the context dependence by focusing on the features of the object that will be the same no matter where we observe it. In this scnsc wc do not carc what function or meaning the DNA has, only that DNA requires something to already exist that can build it (e.g., a cell, or a laboratory)—that is, we care that DNA is itself a product of evolution. DNA is evidence of something that carried function or meaning, because it could not exist without it. We cannot look at DNA on its own as an object and say what its meaning is. But we can measure the minimal path to make DNA, independent of what built it or where we find it.

This brings us to one of the most important reasons we use the minimal path in assembly theory: it is agnostic and does not depend on the chemistry of life as we know it. We can go in the lab and measure it using several different instruments, and we can measure it for *any molecule.*

But we can do this only if the molecule exists in many copies, a key feature of objects in assembly theory. There are, in fact, only four important numbers to understand in assembly theory: 0, 1,

2, and many. If an object never materializes (copy number = 0) then it does not exist. Current physics allows for the possibility that a complex object could form by random chance, but these objects cannot persist or be reproduced and be found in high copy number. So there is a possibility of novelty but no selection on that novelty. In assembly theory, objects can materialize only when a lineage of other objects have already been selected to exist that contain knowledge of how to make that object. This lineage can make the object for the first time (novelty), and because the lineage itself is already selected to persist, it could make the object again (selection). This is very different than saying the object could spontaneously fluctuate into existence, because a view of the object and its lineage as the same thing allows the object and its history to be selected together, and therefore for the object to persist and be reproducible once discovered.

Thus, if an object materializes with at least two copies, this indicates there must be a reliable mechanism in the universe to produce that object. And if an object materializes in many copies, it is even stronger evidence there must be a reliable (selected) mechanism that is also material and includes memory of how to make that object. Thus, for a complex object to exist in many copies, there must be many copies of other objects that can make it, and many copies of objects that make those objects, and so on, forming a chain of objects that can cause others to come into existence. You might think about it as a stack of objects building other objects—each layer in the stack must itself be selected before it can build the next layers. In fact, this is how we conceptualize the entire evolutionary history of life on Earth in assembly theory. It's objects making objects all the way down, or as Lee likes to say, "life is the universe making a memory."

Consider Thomas Edison's invention of the light bulb. He went through approximately one hundred different designs, slightly modifying them and selecting on the things he thought were working, before he invented a functional bulb. Because the bulb worked, it was worth reproducing, and physical systems (us, our technology) that already existed invested energy and resources in producing more. It is the functional bulb that was the one manufactured in factories. Now light bulbs exist in the millions if not billions of copies around the world, evidence that information to assemble them (a memory of the steps) was selected to exist. These need not be exact copies: there are many variations of light bulbs, like LEDs, floodlights, and microwave bulbs. The key point stands: multiple copies of very similar objects imply that there is a reliable—selected—process that exists for generating that design and its variations. Light bulbs would not exist on Earth outside of evolutionary selection.

This does not mean that light bulbs are the only possible solution to the problem Edison solved. There are many examples in biology and technology in which independent series of selection events will produce different functional solutions to similar evolutionary problems. An example is the case of invention of similar scientific theories at the same time, such as how the Heisenberg and Schrödinger versions of quantum mechanics were developed simultaneously and only later found to be functionally equivalent theories. Another is how marsupials evolved along a different trajectory to have external pouches instead of internal placentas to gestate their young, but both solve a similar problem of requiring extended care during the earliest stages of development. These cases demonstrate the path dependence inherent in any evolutionary process: the features historically contingent within the lineage that constructs a given object do matter.

Assembled, Not Random: Why Evolution Is Necessary to Build Complexity in the Universe

We've touched on how the laws of physics are currently written in the form of initial states and laws of motion. The ontological status of both of these places them as external to the universe they describe, that is, they are boundary conditions for describing what happens. In fact, it is this feature of current theoretical physics that leaves room for philosophical interpretations like the simulation argument and intelligent design, which both posit in different ways how the reality we live in was designed (either by a programmer or a God). For us to be inside a simulation requires a programmer external to our universe and a code or program that when executed runs our universe. This is consistent with intelligent design, which would posit a very similar explanation for reality by replacing "programmer" with "designer": a designer of the universe exists (perhaps a God) who specified the initial conditions from which our universe was born. In cases where the programmer or designer does not intervene, this is also, perhaps shockingly, not very different from how modern physics views the universe. In our current conception, laws must exist that are autonomous to our universe (that is, they are not emergent properties within the universe), and an initial condition subject to those laws generated the universe we find ourselves in. It is the laws and initial condition that form the boundary conditions for the universe, and are purported to explain everything that has happened since the Big Bang. But there is no explanation for where they come from.

In assembly theory, we take seriously that the laws of physics should not contain the design of complex molecules (that would be a very uncreative, preprogrammed universe), and that a high copy number of complex objects therefore implies that a physical system must exist with information specific to each step. Proponents of intelligent design would want this design to come from outside the universe, and current physics pushes this explanation to boundary conditions of the observable universe. But assembly theory aims to tell us how the universe can design and construct itself (with life the best example of that physics). We do so by building backward from what exists now to quantify how much construction (selection) was necessary to get here, telling us something about where we exist locally in the volume of what is possible. Our initial states (elementary building blocks) require less information, and so do our laws (joining operations): in assembly theory nearly all the information about what objects are built over time emerges in the trajectories of lineages.

Paul Davies likes to pose a thought experiment relating to how we can spot evidence of design—not by a deistic intelligent designer, but design by evolution or an intelligent engineer or scientist. The key question is: How do we recognize designed objects? The example he likes to give is to imagine that he is out and about town and sees a car drive by with his birthday on its license plate. He thinks nothing of it, but then sees his mother's name written on the street sign. Next, a cab drives by with a phone number on the side that exactly matches his social security number with the last digit missing, but that missing digit happens to be the cab number. At what point should a set of "coincidences" like this indicate that something fishy is going on? Is it a random sequence of events, or is it evidence of design?

Distinguishing design (via an intelligent agent or an evolutionary process) from randomness is important in the context of detecting alien intelligences in the universe, via the signatures of their technology. Some of the most complex patterns we understand appear entirely random, even if they do contain design (e.g., in cryptography this is an important strategy for safely transmitting messages). So how can we distinguish the artifacts of evolved alien intelligence from randomness?

The problem is that by current measures of what is "complex," randomness can look quite complex, too, even when it is not exactly what we are after. There is a popular set of arguments based on the idea that there exists a "complexity hump"—systems starting from low complexity and low entropy will increase both their complexity and entropy for a time, and then complexity will start to decrease as entropy continues to rise. The coffee and cream model thought up by Scott Aaronson, Sean Carroll, and Lauren Ouellette nicely demonstrates exactly this point.[3] Perhaps you've done the experiment yourself, where you start in a highly ordered, very low entropy state of having your coffee separated from your cream. Next you pour your cream into the coffee and complex swirling patterns emerge, but over time the coffee and cream homogenize into a soft brown color, and you can no longer distinguish the two fluids—your coffee and cream are now in a high entropy state (with low information, because you cannot identify the fluid as coffee or cream any more). This idea of a complexity peak goes back to earlier work in complex systems, including that of physicist Jim Crutchfield and colleagues demonstrating how a peak in complexity somewhere between fully ordered and totally random (high entropy) characterizes the behavior of many systems.[4]

To make the connection between randomness and entropy explicit here, we can return to the definition of entropy we introduced earlier in this book. When discussing Schrödinger's paradox, we described entropy as the degree of disorder. In the case of coffee and cream, the degree of disorder refers to how many ways you can arrange molecules of coffee and molecules of cream. The mixed coffee and cream represents a very large number of possible configurations. I can randomly swap the position (and velocity) of any two molecules of coffee and cream as many times as I want, and I will likely maintain the same mixed state. However, if I took the separated coffee and cream, in the same cup (say, the instant I added the cream when the two fluids were still well separated), and did the same random operation repeatedly, the fluids would not stay in the same separated state but instead would become more mixed over time. This is one way to describe why the mixed coffee is higher entropy and also why we observe that cream and coffee mix over time—more of the total possible configurations of the molecules in the coffee cup correspond to a well-mixed cup of coffee than to one with separated coffee and cream. In physics, the highest entropy state is that with the most possible configurations of its parts: if you randomly pulled a configuration of your system from a hat, e.g., a random arrangement of molecules in the cup, you are most likely to hit what we define as the highest entropy state. This is also why there is an arrow of time toward increasing entropy, because if we sample possible configurations randomly, we are most likely to sample ones from higher entropy states than lower entropy ones, so there is a tendency to flow in the direction of higher entropy each time you sample a new configuration.

The most complex patterns—where cream and coffee are swirling together but not mixed—are considered complex because they are hard for an observer to predict. Each pattern of swirling coffee and cream you might observe is very rare in the space of possible configurations. These complex patterns are also associated with a high degree of information for the same reason: in order to produce a swirl you have to be precise in specifying where you place each molecule—you cannot be very uncertain in those choices. This is not true for the high entropy, well-mixed coffee, where you can put your molecules just about anywhere and expect to still have a soft-brown cup of coffee.

The mixing coffee example gives us a conceptual way to try to understand the difference between complexity and entropy. A measure of complexity that could capture the intermediate state of not fully mixed coffee and cream would have to quantify features of this high information content and distinguish it from how randomness indicates a degree of disorder in the high entropy state. Many of the mathematical measures that complex systems scientists use to do this come from computer science, where the formal language for discussing this feature is called algorithmic compressibility: things that are complex are hard to describe in a compressible way because you cannot reduce the information necessary to specify their structure to something simpler. In the case of the coffee example, we need a more detailed algorithm to put molecules in the right places to generate a swirl pattern rather than a soft-brown one, so presumably the swirl is less algorithmically compressible—another way to say that it requires more information to specify it. The number pi might be considered complex for the same reason, because written out numerically it is nonrepeating and has no regular structure even for the trillions of digits calculated to date. But this example also introduces some of the challenges we face in quantifying complexity. It is traditionally seen as a feature of the system describing the object, not of the object itself, indicating that complexity is in the eye of the beholder (or the machine doing the computation that describes the object). The number pi is not random: there are relatively simple algorithms that can produce it (e.g., taking the ratio of the circumference to the diameter of a circle), and if we have knowledge over the space of all algorithms that can produce pi we would realize that it is very compressible in a particular language (namely that of geometry). But searching over all possible algorithms is not typically computationally feasible. So given a new sequence of numbers, equivalently structureless to pi, it is difficult to tell in most cases if the number is indeed random, with no simpler algorithm that can produce it. In fact, there are entire fields of researchers who work on just this problem.

In assembly theory, we posit that copy number can allow us to distinguish designed or evolved complexity from randomness. Evolution builds on what the universe has assembled in the past. Objects—even the very complex ones—discovered by evolution will reuse parts that were discovered earlier. High copy number is not just a feature of evolved objects, but also their parts. Human bodies are made out of millions of very similar cells and not millions of totally different objects. Large language models are increasingly being used across many other technologies. Across biology and technology we see reuse of components as a necessary feature of building more complex objects. This is not accidental; with assembly theory we conjecture this is, in fact, the only way for the universe to explore the space of complex objects. Complex objects cannot be made unless there was selection and reuse of parts because the space of possibilities is too large to construct these objects *any* other way. This introduces something a bit paradoxical when comparing evolved objects (which reuse parts) to objects with unpredictable properties that are random. Evolutionary systems are less complex than random, because of their reuse of parts. When intelligent species such as ours invent abstractions like computation, these allow formalizing randomness and recognizing where it can be built (e.g., with a computer or a brain). This suggests randomly constructed complex objects, ones that do not reuse any parts but are nonetheless highly assembled, may not just be the product of evolution but also of a level of intelligence that can build objects without the need to reuse parts. Such objects, which must be algorithmically generated, may be our best candidates for technosignatures (signs of technology in the universe).

If we return now to the set of all possible things that can exist, assembly says something interesting. It says that we can describe the abstract combinatorial space of all possible objects as a physical space with just two observable properties: assembly index and copy number. I am identifying this as a physical space because these are the coordinates you need to describe where objects exist in an assembly space, akin to how you might identify where an object exists in a physical space we call space-time by the coordinates of a clock time and a specification of a location in the three x, y, z dimensions that define position relative to some coordinate axes. You can think of assembly index as a depth into the space of what is possible. A larger assembly index means more possibilities at that layer, and correspondingly a smaller fraction of things that the universe can realize with similar assembly. More selection is necessary for any given object at high assembly to come into existence. Copy number captures the resiliency of the physically implemented construction or design process for each thing: it is evidence of the stacks of objects necessary to assemble the object you are observing. The assembly index and copy number give us coordinates among what is possible that tell us how much selection was necessary to observe complex objects.

A consequence of assembly theory is that we should not expect objects with a high assembly number to ever form spontaneously. This, like many features of assembly theory, goes against the grain of standard physics. It is a commonly held belief among many physicists that anything that can possibly exist has a low, but nonetheless finite, probability to fluctuate into existence spontaneously due to random quantum or thermodynamic events. The probability might be lower for more complex objects, but still, the belief is it should be nonzero. The implication is that in that great space of everything that could exist, our universe can randomly incarnate anything into existence, and anywhere. While physicists think that their theories refute intelligent design, the idea of the spontaneous formation of any object anywhere in the universe with no memory required to build it is the greatest argument for it: it implies every point in space and time contains the design of *every* object. This is the impetus behind the famous Boltzmann brain argument. A Boltzmann brain is a fully functioning brain, complete with memories of its entire (fictional) life that arises spontaneously due to the random fluctuations of matter. The probability of a fully formed brain fluctuating into existence in one instant and out the next is miniscule in modern physics, even with conservative cosmological estimates. Yet the mere possibility of Boltzmann brains, no matter how infinitesimal, can and does pose problems for current theories of physics, because the universe is very large, very old, and more importantly will get older and larger still.[5] Given this vast space of chances for such an event to occur, some researchers have suggested that more observers of the universe should be Boltzmann brains—that flash into and out of existence in the blink of an eye complete with a hallucinated reality—than are products of long evolutionary histories emanating from an origin-of-life event, as presumably you and I are. In fact, given that Boltzmann brains are "more likely" than evolved brains, you might even be one! The paradox is you would never know the difference. You could have fluctuated out of existence after just reading the last sentence as you realized you might even be such a brain. But now you are reading this, which means perhaps you only just fluctuated into existence right now and have come complete with false memories of having read the prior sentences.

In assembly theory, Boltzmann brains are not just improbable events in our universe; they are, in fact, *impossible.* To be more precise, they are impossible without running the entire causal chain of events that selected the information for them to be produced in the first place. In short, brains do not exist outside of evolutionary processes that can generate them. They cannot exist outside of bodies (for now these are biological, and I do not preclude the possibility of technological bodies in the future, but the point is, these bodies would also be products of evolution and selection). We hypothesize, based on the predictions of assembly theory, that high assembly objects cannot fluctuate into material existence spontaneously on their own. You are never alone in the assembled universe—every evolved object must come with many others related to it.

Now we just need to determine whether or not we are alone in the universe.

Building a Life Meter

With assembly theory, we are aiming to formalize the difference between inanimate and animate matter. When we do encounter new life, in the lab or on another planet, we want to know it when we see it.

As I've argued throughout this book, where we arrive in understanding life may not look at all like where we started. Abstracting life to such an extent may result in a description that doesn't fit your prior conception of what life is. This is the art of theoretical physics, and indeed science in general.

In biosignature science—the study of the signs of alien life on other worlds—this manner of deep thinking to look for universality in life has not been the standard approach. As I write, many of the leading biosignature candidates considered by astrobiologists for space flight missions do not uniquely distinguish life. This is because they do not rely on a hypothesis about what life is, but rather on analogies to what life on Earth has produced. Some of these leading biosignature candidates are things like atmospheric molecular oxygen (O_2) on an exoplanet; amino acids on Mars or in meteorites; isotope fractionation in rocks in the solar system; or even the physical shape, also known as morphology, of fossilized objects.

Abundant O_2 in the atmosphere of an exoplanet is considered a candidate biosignature because our own Earth's atmosphere has abundant O_2, produced by the photosynthetic activity of living organisms.

Amino acids are considered a candidate biosignature because they are an important component of the universal biochemistry found here on Earth: all organisms use roughly the same set of twenty amino acids for building proteins.

Isotope fractionation is the separation of isotopes of the same element (atoms sharing the same number of protons but differing numbers of neutrons) in a physical system. Biology displays distinct patterns in the relative ratios of carbon isotopes, which are regarded as characteristic of metabolic pathways evolved on Earth. This has led to proposals of isotopic fractionation as yet another candidate biosignature.

Morphology is a biosignature if we identify "lifelike" formations in rocks of alien origin, the premise being that we might identify some shapes as deriving only from microbial activity and not being of so-called natural origin.

All of these biosignatures are subject to false positives—

nonbiological processes that can mimic the same signatures as biological ones. Consider a planet devoid of life but with abundant water on its surface orbiting a red dwarf star. This planet would produce an O_2-rich atmosphere via photolysis (the breaking apart of water by light) with no photosynthesis. There are potentially thousands of different kinds of amino acid molecules (not just the twenty used by life on Earth), and a score of these are found in abiotic settings, including meteorite samples, suggesting many are readily formed in the absence of life. Deciding which amino acids could be produced only by life would require a deeper theory than mere heuristics can provide. There are many geochemical processes that lead to isotope fractionation. These challenge our ability to assign specific patterns as unique to biology because, again, we would need a deeper reason for making such assessments. Morphology is challenged by the fact that many geochemical processes can also make structures that appear as cell-like fossils would.

The trouble with recognizing alien life is that we do not know what it means to be alien, or to be life.

Most of the excitement around what we are doing with assembly theory has so far focused on how it provides the first theoretically motivated and empirically tested solution to the problem of ruling out false positives in alien life detection.

Lee and I first met at a NASA-sponsored workshop in Washington, D.C., on the topic of alternative chemistries for life. I was a NASA astrobiology postdoctoral fellow at the time and was part of the organizing committee that had gathered approximately one hundred scientists together to weigh in on the topic of what chemistries for life might be possible. In a conversation with Andy Ellington, whom we met at the very beginning of the book, I challenged the idea that getting a Darwinian replicator (a molecule or set of molecules capable of Darwinian evolution) was sufficient to solve the problem of the origin of life. Andy mentioned in his lecture how I had disagreed with him on this point. Shortly after Andy's talk, Lee approached me with a very direct question: "What do you mean a Darwinian replicator is not enough?" We debated and it became clear that although we did not exactly agree, we both saw the same problems with current approaches to the origins of life. Over many conversations, debates, and even arguments over the years about what life could be, Lee and I converged on the ideas of assembly theory. Lee was always adamant that whatever ideas we come up with must be testable in the lab, so we had to tie them to concrete measurements we can do. And he was adamant that the theory should explain how it is that only evolution produces "complex" objects.* I was adamant that whatever physics governed life would have to explain the features of reality we call "information" and "causation" and how these are the dominant physics in what we call living systems.[6] We both wanted to solve the origin-of-life problem. But we realized to do it we would need both a theory that predicts when it should happen, and a way to test and measure if we succeeded. This meant we needed a way to detect life, even if it is instantiated in alternative and potential radically different chemistries. We needed to

*I am putting "complex" in quotation marks because assembly theory has a different sense of what constitutes complexity than theories that came before. Most theories attempting to quantify it assume unpredictable objects are complex, but assembly theory says the most assembled objects are those that have a high assembly index and copy number, and these come into existence only by reusing parts, which means they are not maximally unpredictable.

build a "life meter," as my colleague Paul Davies would say: a way to recognize life as we don't know it, and even as we can't anticipate it.

A good life meter should be capable of identifying things that cannot be produced in the absence of life, ever. There should be no false positives within the limits of measurement error, because if there are we have not isolated the physics uniquely explanatory of life. If there are false positives for our life measure, then we have not succeeded in identifying life as a natural kind but instead are still treating it as a subjective category. Conversely, if we do succeed in ruling out the possibility of false positives, we may land on a concept of life that does not conform to our a priori, subjective expectations.

Many people want to detect alien life for its own sake. I want to go through the process of aiming to detect it (and hopefully succeeding if it is out there) so we can better learn *what life is.* Predicting and then confirming through observation or experiment the existence of alien life is the ultimate test of our theories about the fundamental nature of life. To validate a theory we need only detect one example of alien life, so if we can't find it out there, the best way to discover it might just be to do an experiment to generate alien life right here on Earth. To do either, we must be able to identify when observables unique to life first emerge from abiotic processes—in other words, we must be able to distinguish when nonlife transitions to life. If we can do that, we can find aliens in an experiment on Earth or on another planetary body. My use of *alien* here is probably a bit alien: how can we discover aliens on Earth and why do we need to know what life is first? Let me clarify. Currently we use the term "alien" to describe almost anything unknown, from an unexplained feature in the spectral lines of an exoplanet atmosphere to unidentified aerial phenomena (UAP) in pop culture. These are not alien; they are unknown. What will really be alien are examples of life (biological or technological) that have traversed a completely different evolutionary trajectory than we have. If we make an example of that in the lab, or discover it on another world, we will learn what "alien" really is. In contradistinction to most popular narratives of alien contact that treat aliens as mysterious, making first contact (for real) with aliens will happen precisely when they are no longer unknown and mysterious.

In assembly theory, definitive biosignatures are configurations of matter with high assembly because these can form only as products of living and evolved systems. High assembly occurs only when you have objects with high assembly indices *and* high copy numbers. It is not enough to have objects that could in principle be produced by evolution; you must have many of them to indicate they are selected. By our reasoning in assembly theory, life is the only mechanism in the universe that can produce high assembly configurations of matter. This means we need to define a threshold assembly index and copy number that determines when objects *must* have been produced by a causal chain of stacked objects, meaning life. We could then define alien rigorously as causal chains that have a different route emerging from low assembly chemistry than we do (i.e., their lineages have traversed the chemical universe along different paths).

With a rigorous theoretical definition of what it is to be alien and a way to measure it, we can go out in the universe and look for it. Ultimately our theories and conjectures about life should relate back to empirical observables, otherwise we cannot prove them. So Lee and his team decided to test the idea of a threshold in assembly. It should occur if there is an assembly index and copy number above which we find only objects that we know are produced by life.

To make an alien life-detection system, the first step was to confirm that assembly index can be measured with laboratory instrumentation. This is where mass spectrometry comes in. A mass spectrometer is a common piece of lab equipment in chemistry labs, often referred to simply as a mass spec. It works by disassembling molecules into fragments. I once had a lesson from Emma Carrick on this in Lee's lab—she explained how a mass spec works by breaking a molecule apart into fragments, and then breaking those fragments into parts again. She can detect each fragment as an individual line, or feature, in the spectrum the molecule produces. By breaking apart many copies of the same molecule, mass spectrometists can explore all different parts of the molecule. To calculate the assembly index, we therefore need only determine the minimal set of parts necessary to reconstruct the molecule—the size of that set is the length of the shortest path. Emma was using a very fancy piece of equipment called an Orbitrap to do this. The futuristic-looking Orbitrap stood in stark contrast to the historic city of Glasgow visible through the window behind it. Each peak she clicked would run fragments through the Orbitrap again to resolve more features of their substructure. Seeing this made it feel more intuitive how the mass spec is able to pick up features of the causal structure of molecules as captured by their assembly space. The device allows you to recursively take a molecule apart. It's a bit like if you took a soft hammer to our LEGO

Hogwarts to break it into pieces, and you did that again and again to get down to individual LEGO blocks. You could record all the pieces you got along the way, and if you had a second LEGO Hogwarts you could do this again and get a larger sample of the possible subparts of the structure. Doing this repeatedly therefore starts to reveal more and more features in the object's assembly space.

With wide eyes, Emma told me how a more typical use of a mass spec is to identify molecules by comparing their patterns of peaks against known samples. Her voice rose with excitement when she relayed how she was instead using it to develop a method to identify whether molecules were of alien origin or not.

What Lee's lab found is the assembly index of a molcculc correlates directly with features of the fragmentation pattern of the molecule. By breaking the molecule apart into fragments we can look at the number of unique parts, and this is how we determine the assembly index as the minimal number of parts needed to build the molecule. A mass spectrometer requires a minimum ten thousand copies of the molecule to resolve this structure. Therefore, *any* measurement made by a mass spec automatically accounts for our copy number requirement.

I made the argument already that we should consider the assembly index an *intrinsic* property of a molecule. This is, in fact, one of the boldest claims of assembly theory and one I view as essential to using the theory for solving the origin of life. But what is the empirical evidence? Some have argued that assembly index is not intrinsic because it is just picking up on how a mass spec breaks molecules apart. So Lee and his team set out to demonstrate it could be measured other ways. They have confirmed that you can measure assembly index by at least two other techniques: infrared spectroscopy and nuclear magnetic resonance. These methods probe the structure of molecules in very different ways, yet nonetheless they all give highly correlated measurements of the assembly index of a molecule. This indicates the assembly index is a feature of a molecule that we can find via a multimodal suite of different measuring techniques, suggesting it is indeed a feature of the molecule and not strictly the measuring device interacting with it.

To prove assembly theory could work as the foundation of a method for life detection, the next step was to take samples produced from nonliving and living systems to determine whether there is a threshold in assembly index above which we find only molecules derived from the living samples, and none from the nonliving ones. What we observed experimentally is that this threshold in chemical space for observing objects made by life (and only life) is fifteen steps (for the chemical space on Earth at least).[7] Why fifteen? We had already predicted this based on theoretical grounds alone, so we can gain some intuition based on the theoretical argument. Consider again how we built up an assembly space by taking two parts and joining them together. Now imagine you are doing that but you have no specific target object in mind. You just randomly join objects from your assembly pool. At each step, there is a more than exponential growth in the number of things you could have built, but there is only one you did. Now imagine I do the exercise, too, with no knowledge of what you built. In fact, we ask everyone else on Earth to do the same. For objects with only a few steps many of us would produce the same structure, even if we do so at random. But as we increase the number of steps it is increasingly less likely that any two of us would produce the same exact structure. If we increase the steps further it is increasingly less likely that among all eight billion of us we built just two things that look at all the same.

Chemists measure amounts of molecules in moles, where a mole is 6.23×10^{23} molecules. This is pretty big. One cup of water holds approximately 13 to 14 moles of molecules, or about 10^{24} molecules. This is roughly the same order of magnitude as estimates of the number of stars in the universe. Using the same logic as our random structure exploration, Cole Mathis, then a postdoc working with Lee, now starting up his own lab at Arizona State University, was able to calculate that at approximately fifteen steps into the assembly space of molecules, we should not expect to see more than one copy of a given molecular structure in a 10^{23} molecule sample. In other words, fifteen is about where we hit the threshold of one copy per mole without selection. We cannot measure molecules at this level of sparsity. (Imagine trying to target just one odd star in the entire universe!) Chances are such a rare (low copy number) molecule will never be identified unless something constrains the construction process not to be random, that is, if selection is at work and more copies are made. It is important to note the threshold of ~15 derives from experimental data, but the calculations help us gain intuition for why we observe what we do. Cole was co–first author of the paper that presented the concrete empirical evidence in support of this theory. He loves the origin-of-life problem as much as anyone I have met, and he is motivated, along with many of our other brightest early-career scientists in this area, to help organize new modes of science to enable us to solve it.[8] Cole's calculation was important and subtle because it demonstrates how it is impossible to *observe* molecules with an assembly index higher than the threshold in the absence of a directed process with knowledge of how to produce that molecule. The directed processes that allow objects to exist beyond this threshold can emerge only through evolution and selection, and can include things like us (humans or AI) designing a molecule or a cell whose metabolism was selected to construct it. Work being done in my lab, as I write, aims to refine the theory around this threshold as a phase transition in the possibility of objects existing. You can think of it a bit like phase transitions in the thermodynamic theories of matter, like when ice transitions to liquid, but here the phase transition is between abiotic objects and life. This is worked out by Daniel Czegel, Gage Seibert, and Swanand Khanapurkar, who are demonstrating more rigorously why all assembly spaces should have a clearly defined threshold above which only objects produced by life can exist in measurable abundance.

With the prediction that we should not ever observe molecules above the assembly threshold in high copy number outside of life, Lee's lab went and tested it in the laboratory. Taking a variety of samples (including some Scottish whiskey!) from living things, dead things, abiotic things, and samples blinded by NASA, the mass spec measurements verified that the only empirical samples with assembly index >15 were derived from life. What this suggests is that there is an assembly threshold in chemical space that nothing but "life" (or whatever physics governs the phenomenon we call life) can cross.

The combination of theory and experiment allowed us to uncover how assembly can be viewed as an intrinsic property of evolved chemical matter. Because it is measurable, it can be used for alien life detection. By having a theory that is built starting from what we can measure, we are now looking at life—alien or us—in an entirely new way that forces us to reinvent the categories of "life" and "alive" to ones that might actually be natural kinds. And just as we have with the invention of every new field of physics, in the process we may need to reinvent other pretty fundamental concepts that we thought we had a grip on, like the nature of time and matter.

Time as Material

"We are living in a material world. And I am a material girl." Such is the brilliant insight of singer/songwriter Madonna in her 1980s hit "Material Girl." In the philosophy of science, materialism is the belief that nothing exists except matter and its movements and modifications. According to a materialist, physicists are those who study what really exists, because physicists are the ones who get to define what "matter" is in the first place.

What if we don't have the definition of matter right? We saw in talking about the hard problem of matter why it is so difficult for this very basic, yet fundamental, concept to be clearly pinned down. We may be premature in assuming our current understanding of matter is sufficient. It was developed to study tiny things (elementary particles, in some cases molecules), but this is not the material world Madonna was referring to (which she intended to point to social reality), and it describes little about the material world you and I both live in. The vitalists did not succeed in explaining life because they could not do so in the absence of a material narrative. The reductionists did not succeed because current definitions of matter as elementary particles, atoms, and molecules do not account for "life" that "emerges" from the interactions of many, many molecules.

Physicist Sir Roger Penrose, a Nobel laureate for his work on black holes, has proposed some provocative ideas about material reality, stemming from his nonconventional thinking on the relationship between quantum physics and minds, which he thinks must be unified in a theory deeper than quantum mechanics. His take on materialism is to quip, "I don't like the word 'materialist' because it suggests we know what the material is."[9]

Most people who are trying to reinvent physics from the bottom up work on what are called unified theories in physics, that is, they work on the unification of quantum mechanics and general relativity (our current best theory of gravity). Physicists want to merge our existing theories of reality because we want fewer equations and ideally simpler descriptions of nature. However, this concept of unification, by combining existing theories to find one that is more fundamental, is a relatively new phenomenon in the history of physics. Over the longer-term history of human thought, unifications were typically driven by open existential questions and novel but deep ideas that might help us connect things we did not previously realize were deeply connected.[10] They were driven by frontier questions, not a desire for mathematical simplicity.

A prominent example is the unification of terrestrial and celestial motion, which occurred in the 1600s with the work of

Galileo Galilei and Isaac Newton. In the words of Isaac Asimov, the prolific science fiction writer and scientist, "We all know we fall. Newton's discovery was that the moon falls, too—and by the same rule we do."[11] Galileo tracked the motions of balls down inclined planes and studied the orbits of the planets and moons through the newly invented telescope. Isaac thought through a process of abstraction in a way that no other mind had done before—writing down concise mathematical laws that described the motions Galileo Galilei, Johannes Kepler, and others had studied before him. After centuries of human thought that would suggest otherwise, Earth and space were discovered to be governed by the same laws.

In the last century alone, we saw three major unifications in our understanding of the world. In the early 1900s Albert Einstein, by studying the properties of light, was able to unify concepts of space and time within a single mathematical structure. In his invention of space-time, he gave curved geometry as the explanation for gravitational force and the behavior of accelerated bodies, and he made possible the prediction of phenomena we had not previously imagined, like gravitational waves (ripples in space-time caused by massive objects) and gravitational lensing (the bending of light due to the curvature of space-time near massive objects). The early 1900s also saw the unification of wave and particle descriptions of matter. Key to this was Albert's theory of the photoelectric effect, which posited light must travel in discrete packets, or particles. He won the Nobel Prize for this work. Louis Victor Pierre de Broglie, a physicist as well as the seventh duc de Broglie, expanded the idea of a wave-particle duality beyond describing just light to also include matter when, in his 1924

PhD thesis, he proposed that electrons can also behave like waves. These ideas laid the foundations of the theory of quantum physics.

The final unification of the 1900s is one that may seem to have little to do with physics. Yet we may find it has everything to do with physics in subsequent centuries: this is the unification of reasoning and calculation. George Boole played a pivotal role in this unification with his book *The Laws of Thought*, published in 1854, which outlined how to do algebra when the variables are not numbers but instead are truth variables (e.g., "true" or "false"), providing a formalism for describing logic operations. His work laid the foundations for the invention of computation as a paradigm for regularizing how we humans think into a mathematical structure. In the 1930s Alan Turing took this further than anyone before him by developing the conceptual foundations of automated thinking machines.

Thus, the unifications of the last century were separately about:

Space-time, unifying space and time
Matter, unifying waves and particles
Computation, unifying logic and calculation

At the beginning of the twenty-first century, we still have not unified computation with our other concepts in physics. It stands apart from physics, although many have speculated that the laws of computation should be considered as laws of physics.[12]

You may have noticed a pattern across the different unifications we just discussed. What we thought after a unification was almost universally different than what we thought before. Consider yet another unification from the history of physics: that of electricity and magnetism, which occurred in the nineteenth century with the work of James Clerk Maxwell. Magnetism arises in the interaction of some materials via their repulsion or attraction. Electricity, by contrast, is the flow of currents of charges found in lightning or in electric circuits. These were thought to be entirely distinct phenomenon before James. By unifying these, James gave us a theory of electromagnetism, which describes the behavior of light and radiation together as manifestations of the same thing. After this unification, we could understand light as the oscillation of magnetic and electric fields; this idea had been completely absent before their unification. Using the language of Thomas Kuhn, it was a paradigm shift in our cultural understanding of electricity and magnetism.

Assembly theory represents a new possible kind of unification: that of matter and computation. In the process of this unification it also forces us to reinvent some things. Standard physics treats time merely as a backdrop that objects move through, whereas in assembly physics objects have a size in time (quantified as the assembly index). As with all other unifications, when we start to see how things look the same, it carries the caveat that we will think about those things entirely differently after a unification than before.

Computation allowed us to describe mathematical operations algorithmically, that is, in a finite series of steps. An example is adding one to any natural number: input 6 and you get 7, input 9 and you get 10, input 39 and you get 40 . . . you get the idea. We can automate this process by writing a description of how to map any input to an output. That's an algorithm—in this case, the algorithm is: add one to a natural number, repeat. A key feature of computation is that its properties do not wholly depend on the machine performing it. The same physical machines can run different algorithms, and the same algorithm can be run on different machines. The details of the precise implementation might be different; for example, when you run Microsoft Word on a Windows machine versus a Mac you may see different features, but the overall functionality is the same—what is sometimes referred to as the multiple realizability of software, because it can be run on different hardware.[13] Nonetheless, what computations we can do is dependent on the laws of physics, because computation must occur within a physical system—what is called hardware in computer science.

Computer hardware is made of matter. Matter is currently defined as physical substances that occupy space and have mass. Light particles—photons—are not matter because they do not have mass. So we might more rightly say physical stuff is made of matter and light and therefore computations must be made of matter and light. Matter has so far been defined by a small set of properties, such as mass, charge, and spin. The most elementary objects in our universe—the quarks and leptons of the Standard Model of Particle Physics—enter our theories as point particles. This means we do not think they have any internal structure, and in fact they do not have a spatial size that we can measure. Some suspect these elementary particles might have substructure, e.g., if string theory is correct, the fundamental particles of the Standard Model that make all matter would themselves be made out of vibrating strings. Identifying any such substructure empirically has so far defied our efforts for experimental validation.

Treating objects as points shows up in other areas of physics too. In Newton's laws of gravitation, and even for many of the applications of Einstein's, we theorize that the gravitational force cares only about mass and relative motion, so we describe mathematically the objects in these theories as points and assign masses, positions, and velocities to these point-like objects. If you want to describe the motions of the Earth and the Sun you can do so by treating each as a mathematical point orbiting the other and nonetheless retain incredible predictive power about each of their motions.

We've already started to see how assembly theory defines objects differently than we have in more traditional physical theories. In assembly theory, objects are described as assembly spaces that constitute the paths to build them. In assembly theory we build up objects algorithmically, via a finite series of mechanical steps. The operational steps in this algorithm cannot just be any we could run in a simulation on a computer; they must be steps that are physically implementable in the real universe. Thus, for making molecules, the set of all such "computations" is restricted to all bond-making operations consistent with the rules of chemistry. The assembly index of an object reflects the minimal history of physically possible, recursive operations that must occur for an object to appear in the universe.

How assembled an object (or set of objects) is, quantified by its assembly index, can be interpreted as the size of the object in time. Here I am not talking about time as the ticking of a clock, but instead I am referring to a causal structure concept of time, quantified as the number of steps to build the object (e.g., in terms of the causation necessary to generate the object). In molecular assembly theory, we interpret the assembly space as the physical molecule, meaning that the material description of a molecule is composed of the physical rules—or algorithms—for its construction. If you want to describe a molecule's behavior in terms of evolution and selection—e.g., where that object sits in the space of what is possible for the universe to construct—then its assembly space is the relevant physical representation of the molecule.

I'll note there are many ideas in computability theory—the study of algorithms and their computation—that define complexity in terms of the length of algorithms to perform a given computation. These are not equivalent to assembly index. Epistemologically, assembly is saying something different. It is saying that the construction steps are an *intrinsic* feature of matter, not a property of an abstract machine you might run your computer program on. This feature of matter can be resolved only in how an object is unfolded across time. It is also relatively simple to show that the assembly index has very different properties than measures of algorithmic complexity.

If you hold your LEGO object in your hand, you can feel its weight. Our best theoretical explanation of what you feel is the force of attraction between it and the Earth, where your hand is providing a counterforce that prevents the LEGO from falling. You can observe the yellow and blue colors of your LEGO object too. Our best theoretical explanation of what you are observing in this case is that the photons that originated from reflection off the surface of the LEGO object are hitting your retinas, which are converted to the appearance of the specific colors you see based on the frequencies of the light hitting your eye. You do not see the metric tensor that describes the geometry of space-time, but you can feel what it describes. You do not see the oscillations of the electromagnetic field that describes the energy carried by photons, but your brain constructs colors as a way to consistently map those frequencies to your perception of them.

Assembly theory likewise says something fundamental and very abstract: every object exists in "time" as the aggregation of recursive paths by which the universe can assemble it. Consider this for your LEGO object—each time you deconstruct and reconstruct it, you are unfolding the object across time, resolving features of its assembly space.

By aiming to derive a theory that can explain life—a process of matter constructing itself across time—we come to a new fundamental understanding of how the existence of objects and time are related, but to get there we must redefine what we mean by matter. It is important at this juncture to emphasize the real radical departure here: what assembly theory is telling us is that *complex matter is complex because it has a physical extent not just in space, but in time too.*

Stated another way: the (physical) algorithm is the object. I can explain it better by analogy: it is not possible to fit a Mack truck inside a standard home garage* because the truck is larger spatially than the garage. Likewise, some objects (a complex LEGO structure) are too large in time to come into existence in intervals that are shorter than their minimal size. To evolve stars, planets,

*The technical note is that I am assuming they are in the same relativistic reference frame.

cells, and eventually cities, the universe literally had to get large enough in time to fit these objects, because they are themselves large in time.

The philosopher Timothy Morton has developed the concept of hyperobjects, which they define as objects so massively distributed in space and time as to defy holding a specific location in either space or time, at least from our vantage as objects smaller in both space and time.[14] An example is climate change, which is difficult for humans to conceptualize and therefore mitigate because it exists over much longer timescales than us as individuals, both spatially and temporally. If assembly theory is on the right track, every evolved object is a hyperobject, with a well-defined depth in time, measured in terms of the minimal route the universe could take to produce that object. Objects can come into existence only when sufficient evolutionary time has passed for that object to fit in the stack of co-constructing objects.

There is an intriguing philosophical interpretation of this, which currently is among my favorite ideas to play with. The foregoing suggests that features of reality appear "material" or "physical" to us if they are smaller in time than we are. Examples include elementary particles, atoms, and biological cells. By contrast, objects that are larger than us in time look "abstract" and "informational." An example is human culture as mediated by our languages, which does not look like a tangible physical structure to us as individuals, yet nevertheless binds us together as a collective object that undergoes selection and evolution. Artificial intelligences like ChatGPT are alienating to us because they take structures that are large in time—things like language distributed over many human minds—and compress them to an object on a size and timescale we can interact with (the large language model).

In assembly theory, what we call life and concepts of computation, matter, and time are all the same thing; the physical structure underlying them is the assembly space. We can now unify the concepts of computation and matter, but we will need to redefine what we consider material. In assembly theory, the material property is the assembly space, which has a size or depth in time. The size of an object in time, in terms of how much "information" or "causation" is necessary to build an object, is a material property.

In order for life to exist as a natural kind, we need a new concept of time as material.

Life but Not Alive

My friend and colleague Michael Lachmann is the only person I know who will cite just how old he really is when asked his age: 3.8 billion years, give or take a few. He'd say the same about you and me. Of course, when Michael makes this claim, he is not suggesting that the atoms and molecules he is made of are that old. What he's really referring to is what he calls his lineage of propagating information. Ribosomes are a good example. These large macromolecules are the part of cellular biochemistry that translates the information encoded in your genome into the proteins that mediate the functions in your cells. Ribosomes are present in all cellular life we know of and are believed to have been present in the earliest known life on Earth. They hold the distinction of being among the oldest physical structures on the planet, even older than most rocks. Ribosomes must be continually rebuilt by cells because their half-life, the time you would expect half of a sample to degrade, is only five to ten days. But if they last only a few days, how can I claim ribosomes are older than most rocks? It is because ribosomes are also an informational pattern propagating on our planet, which we observe in the construction of individual ribosome molecules. It is the lineage of them being rebuilt again and again over billions of years that is old, not the individual molecules.

Loren Williams, a professor of biochemistry at Georgia Tech, where I did my first postdoc, is an expert on ribosomes and one of the foremost molecular archeologists trying to reconstruct their early history by treating the structure of the ribosome like a fossil. With his lab, Loren developed an "onion" model of ribosomal evolution.[15] It suggests how the ribosome may have evolved in layers over time into its modern forms: these layers can be peeled back much like the layers of an onion to reveal a deep and ancient interior. Loren will note in his lectures that the interior of the ribosome (meaning the bonds and atoms in its deepest interior structure) have not moved in nearly four billion years. Life on Earth has been continually rebuilding these objects with atomic precision since before we are even sure there were oceans. Ribosomes are continually rebuilt within our biosphere because the pattern to assemble them—made physical in other objects—has persisted. We humans have never built a technology that will last this long, but we might. This is the kind of lineage that Michael is referring to as the core feature of what he is.

Michael's point reveals a structure much deeper than what we learn in molecular biology: the molecules we observe in the biosphere are threads through time, keeping memories alive in the present. It is not just ribosomes that are continually reconstructed. Entire cells are reconstructed, and so are you over the course of your lifetime. In fact, *everything* in the biosphere and technosphere must be constructed over time, and therefore all biological and technological objects must be continually reconstructed if they are to persist in time. This means the memory to build these objects must also persist.

The same logic applies to you—you are a temporally extended object. You are not made of the same atoms you were a decade ago, but you are the same person with memories from your past self. You are rebuilding yourself across time—this is why you must continually eat to keep yourself alive.

This brings us to a fundamental observation regarding the physics of life: objects persist in time because they can be rebuilt. Every complex object eventually degrades, so the property that a new copy of an object can be made is a key feature of anything complex you might observe. This property is consistent with assembly theory, and it is one I have already pointed to. Indeed, it is one reason we might describe life as deep stacks of causation of objects making other objects, because for *any* complex object to exist, the memory, or constraints if you prefer, to generate it must also exist embodied in another object.

We have talked about many features of known life: that it is cellular, that it metabolizes, that it has genetic inheritance. Many of these might be common to all things some of us may want to call life. Some of these will turn out to be superficial properties that apply only to life as we already know it. Most attempts to define life have so far focused on individual objects—cells, replicating units, dissipative structures, etc. In developing a theory for life, we're aiming to move past the cursory descriptors to identify the deepest structure that can explain life. In many of our current descriptors we have been missing the critically important detail of history: the individual objects we might associate to life can exist only as part of lineages that extend over time. In assembly theory, these lineages—the ways of building the object—are the object. Lineages are objects and objects are lineages.

A few years ago, before we started to converge on a common language in assembly theory, Michael and I wrote an essay together explaining how focusing on lineages rather than individuals allows us to reconceptualize what life is.

In fact, we used this insight to explain why some things are life but not alive.

Many examples of life can reproduce, but not all. For example, mules and worker bees cannot reproduce, neither can a single astronaut in space, nor a single adult on Earth. Even in vitro fertilization requires multiple people: probably more than "natural" methods because in vitro methods require technological infrastructure and employees of that infrastructure to proceed. Because we can think of so many examples of life that do not reproduce, it is often the case that people want to leave off self-reproduction as a necessary hallmark feature of life. However, it is not so important whether all life can reproduce. What is important is that *all* known examples of life (and all its objects) include somewhere in their history an object that was itself a copy of an earlier object (not necessarily exact). This is in some sense what I mean by memory: in order to make something again, the universe must literally remember how to do it.

By making information—the sequence of operations to build an object—a physical feature of objects, in assembly theory we bring the abstract ideas of computation and construction into the realm of the material. Complex objects, such as molecules, can come into existence only if there is something that can build them reliably, whether it is a cell, an environment, or an intelligent agent. These objects require an algorithmic process to assemble them. Assembly theory considers the algorithm to be an intrinsic property of the object, rather than a feature of the machine that outputs it. To understand the physics of how an object can be made, we must identify which algorithm the universe is using to build the given object. For example, molecules are built by making bonds, which is the foundation of chemistry. We believe this concept can generalize to all objects that can be combinatorially assembled by physical operations in a finite number of steps, and highlights how objects are evidence of the physics of the system that produces them.

The concept of life Michael and I identified is that some objects require information—an algorithm—to make them. These objects will *never* spontaneously form and must always be constructed via selection and evolution. Evolution must select on each object in a chain of causation: assembled objects occur only in stacks of related objects. Lee once described this as how the universe is simpler machines making more complex machines all the way down. We can identify the boundary between a living and a dead universe in the lab experimentally. For molecules on Earth the threshold is at a molecular assembly of fifteen—this is where random, undirected processes fail at producing complex objects in high abundance. You can apply this same concept to the case of the 62.8th trillion digit of π: that number cannot be assembled until there is a physical device (e.g., a computer) with sufficient memory and processing power to perform its computation. Likewise, DNA cannot exist unless there is a physical system (e.g., a cell) with memory of the steps to assemble it. All objects that require information to specify their existence constitute "life." Life is the high-dimensional combinatorial space of what is possible for our universe to build that can be selected to exist as finite, distinguishable physical objects.

Bcing "alivc," by contrast, is thc trajectories traced through that possibility space. The objects that life is made of and that it constructs exist along causal chains extended in time; these lineages of information propagating through matter are what it is to be "alive." Lineages can assemble individual objects, like a computer, a cup, a cellular membrane, or you in this very instant, but it is the temporally extended structure that is alive. Even over your lifetime you are alive because you are constantly reconstructing yourself—what persists is the informational pattern over time, not the matter (at least not in the traditional sense of the word "matter"). Likewise, an astronaut in space is alive, but not in a biological sense of reproducing its lineage in genes. Instead, they are alive in the sense that they can interact with their surroundings to cause things, or to write a book about their experiences that might make it back to Earth and change the trajectories of events due to the minds that read it. The astronaut, even though isolated from biologically reproducing, is still a root for complex causation in our universe, including the reproduction of their thoughts. Being alive is not binary, it is a spectrum. The more things that can be constructed by a given object, the more alive it is.

I used the example of rockets in chapter 2 as evidence of imagination. We can now see how this ties into the same ideas underlying what life is. The idea of rockets existed centuries before they could be assembled as definite, physically distinguishable objects. This is because the things required to build rockets themselves didn't exist yet (e.g., we had a concept of rockets before we had built engines that could power them). Things that require a lot of steps become increasingly improbable to observe if there is no process that has selected for the specific causal chain leading to that object. This is the insight we need to recognize alien life on other worlds, and to evolve *de novo* life from scratch in the lab. In looking for alien life, we are looking for very deep (in time) causal chains as embodied in the objects they generate.

From this vantage, the most alive thing on this planet is our own technosphere, the system we are integrated with that is composed of all technology we have constructed. It is the evolved object with the largest size in time that we know of. It can also cause more things to come into existence than any other structure we know of in the universe. This is eloquently captured in one of my favorite quotes, this one provided by David Deutsch in his book *The Beginning of Infinity*:

> Base metals can be transmuted into gold by stars, and by intelligent beings who understand the processes that power stars, but by nothing else in the universe.[16]

That our technology can capture such a deep regularity of nature and use this knowledge to cause things to happen is a highly nontrivial feature of our universe. Things that are caused to exist because of us include examples from the chapter 2, such as launching satellites to space and synthesizing in high abundance high atomic number elements, among a multitude of seemingly uncountable other things. You are most likely alive if you are reading this, but we should not take for granted that all things that are life are also alive.

Information propagates across time in living matter both over evolutionary lineages and over the lifetime of an individual. Individuals continually construct themselves over a single lifetime by rebuilding all their parts. When the entire individual is copied, we call that reproduction. And that process across longer timescales, both in terms of parts and whole individuals (and variations that arise), leads to evolutionary paths and their bifurcations. If we focus on how objects are made, we can unify the description of individuals, lineages, or even the collection of all lineages from an origin-of-life event: these are all examples of assembling objects contingent on using parts that were assembled in the past.

There is only one example of life on this planet, and it is a bifurcating pattern of information structuring matter. The fundamental unit of life is not the cell, nor the individual, but the lineage of information propagating across space and time. The branching pattern at the tips of this structure is what is alive now, and it is what is constructing the future on this planet.

Four

ALIENS

Individuals in black suits are fighting intergalactic terrorists in a clandestine effort to protect the Earth from those who might seek to destroy it. You are blinded by spotlights as saucer-shaped objects appear out of nowhere and descend into the field where you had just been enjoying a quiet evening in the countryside. Governments organize a coordinated international response to intergalactic visitors whose spaceships have landed in most major cities across the globe. These are just a few of the popular modern narratives of first contact. Each highlights how our visions of aliens are culturally specific and anthropocentric. In our search for life beyond Earth, we cannot seem to get away from envisioning aliens that are somehow like us, both culturally and physiologically.

We cannot imagine the truly alien. This schism between what might be alien and what we can imagine is not just a challenge for science-fiction writers—even we scientists struggle. We anticipate, and indeed organize our scientific expectations around, our belief that we will experience an almost mystical "aha!" moment and immediately recognize alien life when we first encounter it. The guiding aphorism "I'll know it when I see it" is still very much alive in our search for alien life. The truth, however, may be far stranger: we don't know what we are looking for, and we may not be able to recognize or even visibly see, or hear, alien life.

The modern era of the scientific search for alien life began in earnest in 1996, with a very public announcement of first contact. Then-president Bill Clinton stood at a podium on the White House lawn, ready to reveal the discovery of alien life. The discovery was not one of little green men, an intergalactic senate, or squid-like creatures from beyond. Instead, the announced findings, published in the prestigious scientific journal *Science*,[1] were of fossilized microbes in a recovered meteorite thought to have originated from Mars. The first question the press asked President Clinton after he made the announcement was about heated debates on abortion rights—a critically important human issue, but nonetheless evidence of how humanity could not focus its attention on the great beyond for even a few moments, it seemed. Regardless, the event was a pivotal point in how we culturally interact with alien narratives. It marks the first time the scientific search for aliens made it into the international spotlight as a topic of U.S. presidential concern.

The announcement was the first by the White House, and to my knowledge by any world-leading government, indicating that the scientific search for alien life—what happens in academic institutions and research laboratories—is a priority governments should take seriously. Importantly, the Clinton announcement was driven not by sci-fi ideas of what aliens might be, but by a scientific understanding of what aliens could be based on data collected from a rock that originated on Mars. The meteorite in question had been discovered in Allan Hills in Antarctica in 1984 and was named Allan Hills 84001 or ALH84001 for short. ALH84001 was ejected from Mars—likely in a meteoritic impact—and its trajectory through space happened to intersect the Earth's path such that it impacted our surface. In 1996 a team of researchers studying the recovered meteorite sample discovered what they thought were microscopic fossils of bacteria embedded in the rock. They attempted to rule out the possibility these microstructures could be from Earth, and they published their findings stating that it appeared the microstructures were native to the Martian rock. The publication of their findings led to an international frenzy, including the announcement by President Clinton.

An important aspect of scientific findings is that they are always subject to being overturned, updated, or otherwise refined. As it happened, the discovery of life in the ALH84001 meteorite did not stand up against the scrutiny of the international science community. Most researchers now agree there are no alien microfossils in the meteorite. Despite much fanfare, we did not make first contact in 1996. This is perhaps just as well. Among all the possible first-contact scenarios we can imagine, meeting dead alien microbes fossilized in a rock is among the less exciting.

While the ALH84001 incident will not be recorded in history as our first intersection with alien life, it was an important milestone. The presidential announcement itself set a new precedent for what first contact might be like. Rather than having aliens invade the Earth in an epic Hollywood-worthy drama, it will be our curiosity and ingenuity as a species that will allow us to discover others "like us." An announcement by a world leader changed the discourse from one of science fiction to the possibility of science fact and sparked a newfound interest in exploring the scientific possibilities.

NASA started investing more concretely in astrobiology research, founding shortly thereafter the NASA Astrobiology Institute (NAI). Over its twenty years in operation between 1998 and 2019, the NAI launched many of astrobiology's most important research agendas and fostered a new generation of scientists tackling the open questions of astrobiology. In 2017, the annual NASA Space Act declared the search for alien life among the primary objectives of NASA for the first time in history. Astrobiology is now a major theme throughout several NASA mission directorates.

The Clinton announcement was by no means the first to make international headlines for the putative discovery of alien life, nor was it the last. Decades earlier, headlines burst with stories of the discovery of little green men. In 1967, Dame Jocelyn Bell Burnell, then a graduate student, detected a strange signal with a radio telescope she helped construct as part of her PhD research. At first she thought it was a "bit of scruff." On further analysis she realized the signal was a series of pulses perfectly spaced at 1.337 seconds apart coming from the same patch of sky. Jocelyn, together with her PhD adviser, Antony Hewish, worked to rule out various sources of interference, including human sources like our own radio signals reflected back at us off the Moon, other radio astronomers' telescopes, television signals, and orbiting satellites.

None could explain the data. After attempts to rule out every known possible human explanation, the regularity and precision of the signal opened speculation it might be of alien origin. In that moment of excitement when our species was on the precipice of anticipating the alien, it was thought that no natural astronomical sources could be so rapid or so precise. We thought, perhaps, the signal could be explained only by an artificial origin.[2]

But we did not discover aliens. This would be a very different book if we had, or it might not exist at all. Jocelyn and Antony discovered a second signal from an entirely different point in the sky with a different frequency than the first. With a second signal, the alien hypothesis suddenly became a less viable explanation; it was deemed unlikely that two different alien intelligences could be attempting to communicate in a similar but not exactly identical way from completely different locations in our night sky (meaning the intelligences would be in entirely different parts of the universe and would be too far apart to even be speaking to each other, so would have to be of entirely independent origin). Soon a third and fourth signal were discovered, also from different locations in the sky, and the alien hypothesis was entirely ruled out because it could not explain the new data. The explanation the researchers landed on was something almost as strange: pulsars. Pulsars are neutron stars, stars whose matter has collapsed under its own gravity, becoming so small and dense that it is all condensed into neutrons. If the material of a star gets any denser than this, the only possibility is to collapse into a black hole. Neutron stars are therefore pretty alien to us, as the densest material in the universe, but they are not life.

The reason we can interpret Jocelyn's observations as the twinkling of pulsars across our sky, and not the twinkling of alien technologies, is because we have a good theory that predicts the signals that Jocelyn observed. Models of neutron stars predict their small size and intensely rapid rotation, and they also predict how the rotation will lead to the emission of beams of electromagnetic radiation from their poles at very regular and very rapid intervals. Because of their rapid rotation, we "see" pulsars as rapidly pulsating bursts of radiation. We have no theory that would explain these signals as originating from alien life.

As I've tried to emphasize throughout this book, what matters most in science is the explanatory power of our theories. The astrobiologist and popular science communicator Carl Sagan is credited with popularizing the saying "Extraordinary claims require extraordinary evidence." This aphorism has, by some, been cast into what is called the Sagan Standard. It is meant to signify how the more exceptional a certain claim is, the higher the standard of proof must be. Claims of discovery of alien life—which would be rather exceptional—therefore should require very strong evidence indeed. However, the Sagan Standard misses the key point about such a discovery and science in general—the proof will rely on the explanation we provide as much as the evidence, perhaps more so. This is one reason UFO sightings have not gained traction as evidence for alien life. It is not merely that the evidence is hard to reproduce, it is that there is no theory for alien life to explain the sightings that itself holds up to scrutiny and has sufficient explanatory power to explain things beyond a few isolated incidents.

In testing the pulsar hypothesis versus the alien hypothesis, we ultimately landed on pulsars as the best explanation not because of the evidence, but because of our ability to explain the observational data in a manner consistent with our current theories of physics, our understanding of cosmology, the limitations of our instrumentation, etc. And we were more confident because of our ability to make predictions about other pulsars not yet discovered: these include predictions like how frequently we should observe them across our night sky, and what range of frequencies we should expect for the observed pulsations. You might recall from chapter 1 that good explanations have broad explanatory reach. There we discussed how better explanations are those that explain more observations, change surprising facts into a matter of course, yield accurate predictions of what one should and should not observe, are falsifiable, rest on relatively few assumptions, and are hard to vary such that changing the details dramatically changes the predictions. UFO theories are easy to vary and can still explain the data: they therefore do not satisfy the criteria for what makes a good theory. UFOs are a good counterexample to what happens in the academy in the scientific search for life, because they rely on extraordinary evidence and very weak explanation. What might really lead us to discover aliens is very ordinary evidence and an extraordinary but strong explanation, one that is hard to vary but fits all the evidence and is consistent with our other theories and observations on how reality works.

In fact, I think we should revise the Sagan Standard entirely if we are to get to a new paradigm in astrobiology. In this new paradigm, we should be able to concretely talk about the possibilities for alien life in a rigorous, hypothesis-driven, and testable way. It is not that extraordinary claims require extraordinary evidence, as Sagan put it, but instead we need to recognize that *extraordinary claims require extraordinary explanations*. The actual empirical evidence can be quite simple.[3] What matters is how the evidence is connected to an underlying explanatory framework and how this interconnects to many of our other observations of the physical world. Scientific revolutions and paradigm shifts are driven by new explanations, not necessarily new evidence. For thousands of years the evidence of planetary motions was on display every night once the Sun had set, but we could not understand those motions, or predict what else could be, until we came up with an explanation in the theory of gravitation. There is a transition in thinking that astrobiologists need to make—we need to care about the explanation for life we are after, and not just sweep the problem of defining life under the rug because it is too hard, or supposedly not relevant. Unless we confront the problem of what life is head-on, we will not be able to discover alien life or solve our own origins. We will not know it when we see it.

The evidence for alien life we often discuss scientifically—whether it is molecular oxygen in the atmosphere of an exoplanet, amino acids on the surface of Mars, phosphine in the atmosphere of Venus, or high assembly molecules in the plume of Saturn's moon Enceladus—is relatively simple. However, the depth and breadth of the explanatory power of using these data as evidence for life varies dramatically. In the case of molecular O_2, phosphine, or amino acids, we are subject to false positives; these can be produced in the absence of life, and indeed their presence is not connected to any deeper explanation of what life is, or predictions of when and where in the universe it will emerge. In the case of molecular assembly, we are testing a theory about what is unique to the physics of life: life is the only mechanism by which high assembly objects can be produced in abundance in our universe. This is because we expect high assembly objects (with high assembly index and copy number) to form only via the processes of evolution and selection. As I will discuss in the next chapter, the same theory holds promise to also allow us to *test and predict* when life originates and what it might be like on other worlds. In assembly theory, life is formalized as a natural kind, and the "kind" we are talking about is the mechanism for how the universe gets complex.

More than providing a philosophical reframing of alien life, assembly theory is important because it offers this radical reframing of how we think about the nature of life in the universe while also being testable. As far as I am aware, it is the first theory in the science of astrobiology to push us beyond the Sagan Standard into territory where the theories are broader in their explanatory power than the immediate evidence we test them against. Whether or not assembly theory proves to be the best explanation for solving the open problems of astrobiology, this paradigm shift to theory-driven approaches is one I hope stays, because it will be critical if we want to solve the dual problems of how life originates and whether alien life exists.

A consequence of recognizing the role of explanation is that first contact is not an event, or even a piece of data. It is a cultural transformation in our understanding of and explanations for what life is. We need to be able to recognize the alien in us before we can recognize the truly alien. As the science-fiction writer Ted Chiang put it to me, the extraordinary evidence that would convince him we discovered aliens is scientific consensus. This is a very high bar indeed! And one worth working toward. It will require piecing together many lines of evidence, but these must ultimately be directed at building an explanation for what life in our universe is and one that we can test empirically.

We therefore need to reimagine what we mean by first contact if we are to make it happen. First contact will not be with a physical artifact but with an explanation—it will be when we achieve a collective understanding of what it is to be alien. It will be when we achieve an understanding of what it is to be one of us: when the idea of what it is to be alien is no longer alien.

Aliens Beyond Imagination

The word "alien" invokes otherworldliness and a vast space of imaginative possibilities for most of us. Even astrobiologists have a hard time anticipating what aliens might really be like. The conceptual artist and experimental philosopher Jonathan Keats pointed out to me the limits of our imagination in a very stark way, by noting how as humans we are hard pressed to understand what it is to be a plant. Despite living all around us, plants are very strange indeed. I recently ran a workshop on the topic of whether plants could compute (anticipating this as a more concrete way of asking if they have any kind of intelligence) and came to deeply appreciate just how "alien" they are. For a plant, which cannot move or verbalize, its behavior is its three-dimensional shape. Imagine if your best way to express your unique history and experiences was to slowly grow the shape of your body to express features of your past experience. All our thoughts about what it is to be alien are so closely centered around ourselves and how we view the world: we cannot really imagine a world where our self-expression must be pushed out through the form of our bodies as we grow in a single place. Plants are just one example that makes clear how the boundary of our imagination does not even intersect with what it is to be among the other multicellular life that surrounds us on this planet.

If we cannot even shift our reference frame enough to understand what it is like to be other inhabitants of our own planet, how could we possibly imagine the truly alien? "Truly alien" here should be understood as other life that does not share any ancestry with our own: that is, that has an entirely unique history with an independent origin. There are no aliens on Earth because as far as we know, all the life we have encountered shares a common history. Even artificial intelligences—sometimes described as alien, are not alien; they are trained on human data, which is itself the product of nearly four billion years of evolution on Earth. AI is as much a part of life on Earth as any of the biological organisms that have evolved here.

How big is the boundary of our imagination, and when might the possibility space we can imagine finally intersect with what aliens could really be?

Assembly theory suggests some surprising new ways of framing the nature of our own imagination with respect to what is alien, because like everything else in biology and technology, imagination should be historically contingent. When we discussed focusing our attention on the hard problem of what exists, we described how imagination could be a consequence of whatever physical mechanism underlies what can be caused to exist. At that point, I quoted Albert Einstein in his observation that "imagination encircles the world." I want to now put this quote in the context of a philosophy derived from assembly theoretic principles, because it challenges a tendency we all share, which is to think that we can really imagine everything. There are commonly held views that point to a misunderstanding of the size of the space of possibilities. Assembly theory reveals the historical contingency in every object, even us. In assembly theory, what we can imagine, like everything else a biosphere generates, is also historically contingent and cannot cover the space of everything. Imagination can get us only to places that are assembled from the combinatorial possibilities existing in what we are now.

In assembly theory we have a concrete and mathematically rigorous way of talking about the space of what is possible.[4] We start always only from what is observable. Given a set of observed structures, we define *assembly observed* (A_O). This is the assembly space of all objects that we can measure. In our LEGO universe example from the last chapter this might include, for example, all mass-produced LEGO castle sets made available since the company was founded in 1949: among these are Hogwarts and the Taj Mahal. We then deconstruct this space to its elementary building blocks. In LEGO these are the bricks. From there we build up all possible structures in an unconstrained way, using all possible permutations of the basic building blocks. This is what we call the *assembly universe* (A_U). It includes objects that can be constructed using the rules of LEGO but also ones that can't be made using well-defined LEGO building rules. For example, it might include structures where two bricks are superglued together, violating the interaction rule of LEGO that blocks must be stuck together using their interlocking structure. The assembly universe does not include the physical constraints of how things can be constructed using implementable operations consistent with what is physically possible, so some things in this combinatorial space simply cannot exist as isolated physical objects, only as possibilities.

Constraining the possibility space to objects that are physically possible to exist (as dictated by the laws of physics) yields what we call the *assembly possible* (A_P). However, not everything within this restricted space is actually possible and certainly not all at once. Assembly possible forms the outer bounds of what is possible for our universe to make, but the full space is itself not physically realizable. It's too big. Even with the approximately 600 billion LEGO bricks produced to date, you could not exhaustively build all possible objects of similar assembly to LEGO Hogwarts or the LEGO Taj Mahal. There are too many to all be realized at once even with this much physical material shaped as LEGO bricks. Constructing possibility spaces that can be physically realized requires an additional constraint: memory and contingency in the construction process of objects. We define this space of possibilities as *assembly contingent* (A_C): it defines the space of possibilities where memory of the past must be selected and those things retained over time literally build the future. It is within the space of assembly contingent that any observed objects—those we would regard as physically real—reside.

A feature of the universe of possible objects, as described by assembly theory, is that $A_O \ll A_C \ll A_P \ll A_U$. That is, we aimed to capture, and quantify, the intuition that what actually exists is an exponentially small fraction of what could exist. In assembly we explain why some things exist as physical objects and not others in terms of the historical contingency inherent in all objects via the structure of their assembly space. Things can come into existence only when the parts to make them already exist. This imposes a locality on existence: objects are more likely to exist together if they share overlapping histories in the objects that came before them (e.g., if they have a shared assembly space).

Recall that assembly theory assumes life is the mechanism of selecting among what can exist, is the only physics that can bring complex objects into existence, and does so in a historically contingent way (e.g., what gets built depends on the past). We should not expect alien life to have the same history we do. Alien life traces out a different trajectory in assembly contingent than we have: objects are defined as alien precisely because their causal history is different from our own. This is a more formal way of saying that aliens have an independent origin and evolutionary history from us.

The assembly universe forms the outer boundary of what can be imagined. We can only imagine this based on what we observe to exist within our own histories. The question of how "alien" alien life really is, therefore, breaks down to whether aliens exist within the same combinatorial possibility space as us or not.

In assembly theory, objects with larger assembly spaces are deeper in time. This means that there are more features in the history of high assembly objects from which the future can be constructed. The realizable future is larger now than it was in the past, and the future increases in size as assembly increases. This suggests there is a horizon or wave front, both in terms of what is imaginable (assembly universe) and—within that—what is physically realizable (assembly contingent), that expands as a biosphere evolves and as it transforms into a technosphere.

When I say aliens are currently beyond our imagination, your own imagination may now be envisioning interdimensional aliens that exist beyond our space-time structure. That is possible. But not practical. In the scientific search for alien life, we need to constrain ourselves to life we might make contact with. If we take historical contingency seriously, we must recognize the limits of our own imaginations as an evolving boundary, but this also means we can evolve the knowledge to imagine what aliens really could be within new explanations we can create.

A question is how deep in time must we be—how large combinatorially must our technosphere be—before we are large enough to share enough features with alien life to know it when we see it.

Anticipating the Alien

A geological map of our own modern Earth includes features like Mount Everest as its tallest peak, the Mariana Trench as its deepest point, and the Grand Canyon as a large crack in its surface rock. Would you ever expect this map to exactly apply to another planet? Imagine trying to use this exact map to predict the surface features of Venus or Mars. This clearly would not get us far in describing the surface properties of these alien worlds. To do that, we need to understand the geophysics underlying planetary surfaces, and indeed this is exactly what planetary scientists do to understand the geology of other worlds.

Yet when astrobiologists approach the question of what biochemistry could be on Mars (or any other world), we do not typically resort to fundamentals. Instead, we do so by direct analogy to the metabolic map of life taken from Earth, a process akin to assuming we can map other worlds by direct analogy to the geological maps of Earth.

A fundamental understanding of life will begin a new era of astrobiology, because it will allow us to move beyond the search for life as one of analogy. It will allow us to search for life as we don't know it. Indeed, it will allow us to search for life as no one knows it. Understanding life in the abstract will allow us to predict what alien examples of life could be like. Our hope with assembly theory is that we will be able to use it to detect alien life, if it is out there, and that even before we make first contact we also might use it to predict features of the possibility spaces aliens might live in.

Assembly theory makes a strong claim: all things life can build are historically contingent. Indeed, we think this contingency also becomes a way to predict novelty because it allows predicting what new forms you might expect to appear based on recombining features of the past.

At the origin of life, the biochemistry that is selected for incorporation into living structures is geochemically contingent. As my geochemistry colleagues at Arizona State University Everett Shock and Hilairy Hartnett advocate, "Biochemistry is what the Earth allows."[5]

Before life emerged on this planet the only chemists were the rocks and water, and as Hilairy likes to point out, the Earth does very different chemistry than human chemists do. What happens in the lab includes only a very restricted set of reactions among what is geochemically possible (and also includes things humans can do that are not geochemically possible).

It is also true that what emerged from our planet as the biochemistry of life is a very small subset of the potential geochemistry Earth could generate. That small subset formed feedbacks that allowed it to become self-sustaining and evolve and increase the assembly of the molecules formed, making possible molecules that the Earth could not produce in the absence of selection on specific histories for their formation. The transition to high assembly objects initiated a cascade of object formation that underlies the evolution of all life on Earth. Our biosphere is an object extended nearly four billion years in time, with a physical size of the entire Earth. Everything that has happened in our biosphere's history is contingent on those first few selected steps. With a technosphere emerging around us now, we see the same kind of contingency and the origin of life happening anew, but in different substrates, continuing the lineage of life from biology into technology. An example within technology is how many modern technologies are historically contingent on the invention of the transistor. Perhaps a different technology could have been invented that was just as useful, and much subsequent technology would instead be scaffolded on this alternative tech. Likewise, could a planet generate an alternative chemical solution for life? And would the subsequent evolution of a biosphere over billions of years be radically different for all time because of a few acts of selection within the space of possibilities very early on, when life got its start?

Recall from our discussions of chemical space just how large the space of possible molecules is, even for relatively small molecules. We estimated approximately 10^{60} possible molecules with molecular weights under 500amu (about the size of a few amino acids). The size of this space is one reason that most prebiotic chemistry leads to a combinatorial mess. Chemists refer to this mess as tar, an undifferentiated mix of molecules, where few individual chemical species are identifiable. This is because no species can be selected in high-enough abundance to be distinguishable from the rest. Tar is what happens in the absence of selection. You get a complicated mixture of low copy number and low assembly number objects.

The most well-known example of tar formation in origin-of-life experiments is the Miller–Urey experiment. This experiment is famous for being the first to demonstrate the production of amino acids under abiotic conditions. If you keep running the Miller–Urey experiment, you end up with a complicated mixture of different carbon compounds that are not individually identifiable. The origin-of-life chemist Steven Benner refers to this as the "tar paradox": if we run experiments aiming to produce the molecules of life, we also get a combinatorial explosion of related compounds that turns into a tar.[6] Tar-like mixtures are also found in meteorites, the most pristine examples of chemistry from the early solar system, which have such combinatorically "complex" chemistry that it is difficult for chemists to identify individual compounds. This kind of complexity is sometimes confused with the complexity that we talk about in assembly theory, but it is not at all the same. The combinatorial complexity of abiotic chemistry is unconstrained, allowing for a large variety of possible low assembly compounds (low assembly index and low copy numbers). It does not make molecules with high assembly (high assembly index and high copy number), or at least there is no evidence of any in experiments or meteorite samples to date. The constructed complexity we talk about in assembly theory relates to how many steps are necessary to produce a molecule or set of molecules and is related to how much evolutionary selection had to occur for that molecule to be possible. It is a complexity that relies on redundancy and modularity of parts that are reused again and again to build up intricate evolutionary structures in high abundance: these are not random, and they are not necessarily complex by standard ideas of what constitutes complexity (because they are distinguishable from random).

The origin-of-life transition occurs when the combinatorial explosion of possible low assembly molecules gets constrained and funneled to select only a subset of possible molecules. Those are scaffolded to build more assembled objects, where those objects in turn build even more assembled ones. It captures the idea that it is objects building slightly more complex objects all the way down. As Lee sometimes says, to solve the origin of life all we need is to generate a simple machine that can build a slightly more complex machine that can build a slightly more complex machine, and so on. Of course Lee does not mean a machine literally, but the analogy is fruitful for visualizing what the origin-of-life cascade could look like.

Now imagine a warm little pond, as Charles Darwin did, and as his grandfather Erasmus alluded to two generations before him. Or better yet, consider all of the microenvironments across a planetary surface as microreactors, and a planet as an engine exploring via these microenvironments the combinatorial space of geochemically possible small molecules.[7] Fluctuations in the density of assembly happen, and life emerges as a cascade of objects building into the high assembly universe. Assembly theory suggests we should not a priori expect life on other planets to start down the same evolutionary paths as life here on Earth. In fact, we might predict biochemistry will be different on different planets and even just how different it is.

An evolutionary process through chemical space is one that selects a subset among the combinatorial explosion and preferentially constructs just those molecules. The constraints that first restricted chemical space in this way were likely mineral catalysts and a cycling of environmental conditions that kept the nascent biochemical system out of equilibrium. Current proposals of what may have driven these out-of-equilibrium conditions include environmental cycling via oscillating wet-dry conditions at the edge of an ocean or pond, or in the cycling of chemistry along the thermal gradient of a hydrothermal vent at the bottom of the ocean.[8] Minerals are known to selectively catalyze some reactions and are widely believed to have played a prominent role in the origin of life. These drivers of disequilibrium therefore play prominent roles in most modern-day prebiotic chemistry experiments.[9]

In the last section we discussed the idea of assembly contingent—the space of all possible historically contingent configurations of objects, i.e., those that can be formed by evolutionary selection and recording of memories of the past. A geochemical soup on a planet, exposed to mineral surfaces and a cycling environment, could potentially start to build along any of these paths, and which path is selected becomes frozen into the rest of the history of life on that planet (if life emerges at all). Small variations in the favored chemistry on a planetary surface can therefore be amplified by evolution and lead to radically different outcomes. Imagine if you decided to travel from the Washington Monument to the Pacific Coast by randomly selecting which road you take at every turn. You can easily see how the paths taken would start to rapidly diverge. The exploration of chemical space in the origin of life is likely similar, although we don't know how many steps will be repeatable from planet to planet. We have some idea that the steps are random—we do not think we live in a universe designed for life to emerge because we don't see life everywhere. Instead, the design of specific objects can be encoded only in other objects.

The challenge is, we do not know if the path life takes on a planet is set before or after cellularity evolves—i.e., before or after there is a planetary transition to a common form of "life" on a planet. On Earth, this happened with the Last Universal Common Ancestor, where early life underwent a transition to solidify a common biochemical architecture across all biology. If the crystallization along a particular path happens before this, then we should expect life on different planets to be chemically different.

Experimentally, we have some indication that we should expect the evolutionary process to play out similarly in bulk features but not in details. Physics constrains the form of wings to be roughly the same across species—there are only so many ways you can grow an apparatus that will let you fly. In chemistry, if you run an origin-of-life experiment in a genesis engine of the kind we will describe in the next chapter and repeat an experiment with the same sequence of mineral additions, or changes in pH or whatever perturbation you like, you'll see the same overall patterns in the chemistry reemerge.[10] Selection is acting on global constraints of what chemistry is possible in that environmental scenario—the chemical equivalent of selecting for a wing for flight. These kinds of experiments indicate that the chemistries will diverge in different environments, just as different limbs evolve under different physical constraints—legs for running, fins for swimming, wings for flying. When we can run origin-of-life experiments at scale, they will allow us to predict how much variation we should expect in different geochemical environments. That is, we will not only be able to speculate that life elsewhere should be different, but we should be able to say how different with quantitative precision.

We can also go into biochemical data to start to identify what features of life on Earth might be generalizable. My lab has been developing what we call a statistical mechanics for biochemistry, where we aim to identify statistical patterns in the chemistry used by life. These might be imposed by features of the environment, in the same manner we analogized to wings and fins. We're looking for patterns that do not depend on the exact molecular composition of life on Earth.

We've discussed a few times already how a major hurdle in our ability to infer universal properties of alien life is that we have only one example of biochemistry to go on—all life on Earth shares the same basic biochemistry. A very different sense of universality, found in physics, could be useful for guiding our search for alternative forms of life. That concept arises from the study of phase transitions. Familiar examples of phase transitions include when ice melts or when liquid water vaporizes. Physicists refer to the physical conditions where a phase transition occurs as a critical point. In the study of phase transitions, an important mathematical tool has been the identification of scaling laws, which allow the quantification of the systematic variation of one property with another—for example, how temperature might scale with pressure. Universality classes in physics are then identified by systems that have the same scaling behavior.[11] Remarkably, this can occur across systems that are completely different in terms of their component parts and microscopic physics: the water-gas phase transition has the same scaling behavior as the one that occurs in magnetic materials between the ordered and disordered phases (where magnets align or point in random directions respectively). Physicists have used this to predict properties of new materials that might be in the same universality class as existing ones—e.g., share the same macroscale properties (scaling) while varying in lower-level microscopic detail (whether the system is made of water or a magnetic material).

This offers the possibility that scaling laws identified for life's chemistry might enable us to predict features of alien chemistry. The first challenge is to determine if generalizable trends indeed do exist. Scaling laws have been discovered across a broad range of biological systems, from how metabolic rate scales with body size to how the number of patents scales with the size of a city.[12] My lab looked at the different kinds of reactions catalyzed by life. These reactions are catalyzed by enzymes—proteins that act as catalysts and accelerate the rate of biochemical reactions. It turns out life as we know it uses just six major classes of enzyme-catalyzed reactions: redox, transferase, hydrolase, lysase, isomerase, and ligase. Redox enzymes are responsible for the transfer of electrons and are typically involved in energy metabolism. Transferase reactions move a part of one molecule to attach it to another. Hydrolase and lysase enzymes break bonds to break molecules apart (hydrolases are a special sub-group that uses water molecules to break bonds). Isomerases change the structure of a molecule without changing the elemental composition of the molecule, and ligases join two molecules by forming a new chemical bond.

Ignoring the details of specific enzyme-catalyzed reactions and instead studying the statistics of the major reaction classes across cataloged life revealed new universal scaling laws.[13] These scaling laws relate the number of each kind of function life uses to the total number of functions found in a given living system. They reveal how biochemical systems organize reaction function in a universal—and therefore predictable—way.

Thus, at least for life on Earth, there are general features of biochemistry, recovered by looking at patterns in the statistics of many biochemical reactions, that might be more universal. The longer-term goal is to be able to map the same kinds of patterns that emerge in our evolutionary abiotic chemistry experiments as what we see in life.

The idea that general features might be universal but the specific implementation will be geochemistry dependent opens new windows into how we view different origin-of-life hypotheses. In chapter 1 we briefly touched on the RNA World hypothesis proposing that the formation of RNA was the critical step in the origin of life. This is one of the leading contenders currently discussed in origin-of-life literature. However, if the foregoing arguments I am making are true, it is possible—likely, even—that RNA is an invention unique to the evolutionary history of life on Earth, and that RNA exists nowhere else in the universe. The things that exist on this planet now are contingent on those first few steps leading us into the path-dependent evolution of a biosphere. I am working on a technical manuscript now with Lee, Michael, and Chris Kempes making just this point. The nucleotides that are strung together to make macromolecules of RNA are adenosine phosphate (AMP), guanosine monophosphate (GMP), uridine monophosphate (UMP), and cytidine monophosphate (CMP) and have molecular assembly index values of 17, 19, 16, and 16, respectively. These are all above our empirically determined threshold value for life, indicating they are products of evolution and selection. Recall Cole Mathis's calculation of an expected estimate of only one copy in a mole of molecules around molecular assembly index of 15. More to the point, the possibility space around RNA is huge, and very likely not to be discovered on the evolutionary path of other biospheres. In fact, we estimate that exhausting the space of molecules with similar assembly to even short sequences of RNA—with only one copy per molecule—on a planet would require sufficient mass that the resources necessary would collapse the planet into a black hole. It is not physically possible for the universe to exhaust all possibilities that could exist on a planet by multiple lines of reasoning. RNA is an invention of this planet and this planet alone.

The argument generalizes. The trajectory of life on Earth as it constructs physical objects from what is possible is probably unique in the entire universe. The vast majority of things found on Earth exist nowhere else, even if there are other inhabited worlds.

Testing the Alien Hypothesis

When I was a PhD student, the holy grail of high-energy particle physics was finding the Higgs boson—the particle conjectured by theoretical physicists to complete the Standard Model. The Higgs was a necessary missing component of particle physics because it would provide the explanation for why some of the other particles have the masses they do. The possibility of a Higgs boson was predicted as early as 1964, in work by Peter Higgs (for whom the particle is named) and others.* After a forty-year search, it was definitively discovered in 2012. That discovery came from the Large Hadron Collider (LHC), a particle physics experiment at the European Organization for Nuclear Research, known as CERN. Located near Geneva, Switzerland, CERN is the largest particle physics laboratory in the world. The multibillion-dollar LHC experiment was built specifically to search for the Higgs boson, because we had a well-motivated theory to guide our search. The experiment would not have failed even if there was no discovery, because it would have disproven an entire set of theories about the properties of the Higgs boson and opened new questions about what other mechanisms could be at play. However, the experiment ultimately succeeded in discovering the Higgs boson because our

*The mechanism for how the Higgs boson can explain mass generation was published nearly simultaneously in 1964 in papers by three groups: Robert Brout and Francois Englert; Peter Higgs; and Gerald Guralink, Carl R. Hagen, and Tom Kibble.

theories were capturing something accurate about how our universe works.

Another major recent theory-driven discovery in physics was the direct detection of gravitational waves in 2017. Guided by Albert Einstein's theory, physicists developed sophisticated computational models for extracting data from very small signals among huge amounts of noise. While gravitational waves were originally predicted by Albert in 1917, it took one hundred years for humanity to confirm their existence.

I am using these examples to highlight how experiments done here on Earth can reveal fundamental features about our universe, relevant to the smallest and largest length scales we can observe in nature. However, to be successful, these experiments must be guided by deep theoretical ideas. Indeed, as I will continue to emphasize, experiments and theory together are what allow us to understand features of our universe that are very distant from our own experience. The Large Hadron Collider is often described as simulating conditions not seen since the Big Bang and is sometimes described as our greatest window onto the conditions of the early universe. Likewise, gravitational wave interferometers provided a new window into astrophysical events that we would not have seen otherwise. They spurred a new field of multimessenger astronomy, which seeks to combine gravitational wave data with more traditional electromagnetic data collected from astronomical sources to reveal new details about astrophysical events not previously accessible.

The relationship between particle physics experiments on Earth and observational cosmology is a fruitful one. I am perhaps biased to use this as a guidepost for what astrobiology could look like because of my training in particle physics and cosmology. But it motivates me to ask whether those interested in discovering alien life and those working on origins of life could similarly see fruitful collaborations that enable new discoveries?[14]

When we started out to develop assembly theory our goal was (and still is!) to solve the origin of life. But the theory is first proving itself most useful as an agnostic life-detection method, perhaps the first driven by a robustly falsifiable hypothesis. The connection for how origins of life and alien life detection might be solved by the same theory is a new development, which to me is promising, because we should expect this only from a sufficiently deep explanation for what life is. If you do not focus on a unifying theory for what life is, it is hard to see the deep connections between origins and aliens. To understand how this constitutes a paradigm shift for our field, it is important to note how for most of the history of astrobiology, the scientific problems of life detection and origins of life have been treated entirely separately.

Here's an explicit and simple example: consider how the origins-of-life and life-detection communities attempt to tackle the problem of homochirality. *Chiral* derives from the Greek word for "hand." Your hands are mirror images of each other, meaning you cannot superimpose one on the other (e.g., if you lay your left hand on top of your right they do not perfectly line up). Chiral molecules have this property of not being superimposable, too, and come in mirror-image forms, called enantiomers, which are often described as left- or right-handed molecules. The vast majority of molecules made up of more than about eight heavy atoms (atoms other than hydrogen) are chiral. Examples in biology of chiral molecules include the amino acids that make up proteins and the sugars ribose and deoxyribose that make up the backbone of RNA and DNA molecules, respectively. An odd—and by odd I mean unexplained—feature of life on Earth is that it has a chiral preference: it uses left-handed amino acids and right-handed sugars. That is, life is what is called homochiral: many molecules life uses are one chiral orientation (left or right) across all known life. We do not know the mechanism by which this property first arose for the life we observe on Earth.

Explaining the origin of homochirality is nontrivial. In fact, I spent my entire PhD thesis working on the problem. In my lab even today we have an effort to reconceptualize the role of chirality in the origin of life using assembly theoretic principles with the hope we might shed new light on what has been a stubborn mystery.

It was in my experiences in communities interested in life's chirality that I first observed the disconnect between the alien-life-detection and origins-of-life communities most starkly. The Astrobiology Science Conference (AbSciCon) is a biannual meeting attracting leading researchers across the diverse areas of astrobiological science. It is a meeting known for the friendships you make (astrobiologists are huggers!) and for the exhaustion of running between so many events happening all at once. Science sessions at AbSciCon cover everything from Mars sample return to the icy moons of the outer solar system, exoplanets, Earth in deep time, technosignatures, the Anthropocene, and geobiochemistry, to name just a few topics. Attending a session on the chemistry of life's origin, you would be likely to see a few talks on the same theme as my dissertation research. This line of research assumes a homochiral set of molecules is a *precondition* for life: almost all origin-of-life scientists focus on how homochirality could arise in abiotic conditions, *before life*. By contrast, if you hopped into a session on life detection, homochirality is put forth as a biosignature—that is, if we send a mission to Mars and find homochiral molecules it is considered a telltale signature that life is already (or once was) present. That one research community could consider homochirality as a prerequisite for life and another should consider it as an unambiguous sign of life is perplexing to say the least. That these two communities should happen to be discussing these ideas in neighboring rooms at the same conference without recognizing the dichotomy demonstrates a lot about how our young field of astrobiology must grow if we are ever to make definitive contact with alien life.

We already start to see connections between the problems of life detection and solving the origin of life with assembly theory—much like how cosmology and particle physics were unified by mathematical theories that explain features of both. Assembly theory was developed with the aim of detecting the origin of life in experimental data, allowing agnostically searching for the emergence of new chemical life-forms. Assembly theory can allow us to understand when matter transitions to life in automated chemistry experiments, because it provides a signature of complex objects that can be produced only by evolutionary systems. This same reasoning is what allows molecular assembly to be a useful biosignature for space missions looking for life on other worlds. *Any* proposal for explaining life should be able to simultaneously allow us to solve the origin of life and to identify alien life on other worlds, because unifying these is how we transition from preparadigmatic science to a new paradigm in the study of what life is.

Alien Worlds

The probability of the origin of life is an important missing parameter in cosmology.[15] Without it we cannot predict the distribution of alien life in the universe.

You might be familiar with the Drake equation, written down by astronomer Frank Drake at an iconic conference at the National Radio Astronomy Observatory on the topic of the search for extraterrestrial intelligence, or SETI for short. Usually this equation comes early in a book about the search for life, but we are not doing things the usual way. Frank was working in the very early days of the scientific search for intelligent alien life. Arguably, modern SETI began in 1959 with the publication of a landmark paper by two physicists, Philip Morrison and Giuseppe Cocconi, who proposed searching for intelligent civilizations via their radio transmissions.[16] They calculated that with modern (1950s) radio technology, it would be possible to detect radio transmissions covering distances comparable to those between the stars. Frank was the first astronomer to actively attempt a SETI experiment. His Project Ozma, conducted in 1960, surveyed just two stars, looking for evidence of radio transmission from intelligent aliens. This was a landmark project and is widely credited with launching the modern era of SETI science.[17]

The Drake equation, proposed by Frank one year after he conducted the first SETI search, was devised to provide an estimate of the number of possible detectable intelligent civilizations in our galaxy. As you can imagine, if you want to design a research program to search for intelligent life, it would be helpful to have an understanding of how many stars we might need to survey to detect anything. From Project Ozma we knew it was more than two, but was it one hundred, one thousand, or even millions? Just how rare *is* life?

Frank's equation broke down the probability for intelligent, radio-transmitting life. He did so by defining discrete, and potentially constrainable factors. Those factors included the number of stars hosting Earth-like planets, which in the 1960s was unknown. Now this parameter is observationally constrained, primarily due to the revolution in exoplanet science that has happened over the last few decades, allowing us to go from zero observationally confirmed planets around other stars to thousands of them. But other parameters in Frank's estimation algorithm remain totally unconstrained. These include all the life terms: the probability that life emerges, the probability of the emergence of complex life and technological life, and the longevity we might expect for a technologically communicating civilization. Estimating these likelihoods is not possible with current data because we have only one data point: our own biosphere's emergence and evolution into a communicating technosphere.

While Frank often argued that the intelligence terms were the least constrained, and also most critical for estimating how many stars we should search, I think we are in fact more constrained by the origin-of-life term. I mentioned that the probability for life to arise from nonlife is an important unconstrained parameter in cosmology. You might wonder why I want to frame it as a problem relevant to cosmology. Well for one, in asking about alien life we are in part asking about the distribution of life in the universe. With current cosmological models, we can explain the distribution of matter in the universe, and we can even predict missing matter—we predicted the existence of dark matter by noticing our models were missing mass necessary to account for the rotation rates of galaxies. We also understand something about the dark-energy density of the universe, something we postulate must be there because we observe the effects of accelerated expansion in the distant universe. While the mechanisms that explain dark matter and dark energy are yet to be determined, we have strong observational evidence for these things. Yet, we have no idea if there is a "dark" sector of currently unknown alien life-forms distributed across the universe because we have no observational evidence they are there.

Some researchers will argue that life must be out there, and that there is no chance we are alone. Our own galaxy is estimated to host two hundred to four hundred billion stars, and there are billions of galaxies in the observable universe. We think most stars have planets, so it follows that there are billions of planets in our universe too. Surely among the billions of possible worlds, some of them host life? The way life arose on Earth so quickly after its formation is often cited as evidence that this should be the case. Earth formed with the rest of the solar system approximately 4.5 billion years ago. We have evidence for life on Earth as early as 3.8 billion years ago, meaning that life formed within a very short window of a few million years after our planet became potentially habitable. Since life emerged quickly on Earth, it is argued, maybe life is likely on Earth-like planets. If you combine this observation with the observation that there are billions of worlds in our universe, then it seems natural to assume we cannot possibly be alone. Even as professional astrobiologists we often fall into the trap of assuming life must be out there because we naïvely want to assume chances are in our favor. Though this reasoning seems natural, it is wrong.

Are the chances in favor of life? This is the part we do not know, nor can we constrain it at present. The cosmologist Brandon Carter is well known for laying out a statistical argument for why our universe seems fine-tuned for life. His anthropic principle argues that if it is possible for universes to exist with other physics—different binding strengths of atomic nuclei, different gravitational constants for how attracted massive bodies are to one another, etc.—then we should find ourselves as living beings only in universes where the physics is suitable for life.[18] Brandon proposed this idea as an explanation for why the constants of nature seem "just right" for us to be here. Tweak them just a little and you would create a universe completely unsuitable for life. His argument rests on the idea of post-selection: by assuming a universe with life exists as the end point, you constrain the possible laws of physics that could support such an outcome.

Brandon applied similar reasoning directly to the problems of astrobiology. While the rapid appearance of life on Earth is often cited as evidence that life should be common, if we post-select on our own existence being contingent on the origin of life happening early, a different story emerges.[19] Of course we could not be here if life did not happen, and there also would be no one to reason about the probability for the origin of life. Properly running the statistics on the likelihood for an origin-of-life event, while incorporating the observational evidence that we would not exist without the one origin-of-life event we know of, leads to a conclusion that the origin of life could be very common or exponentially rare. In fact, it could be so rare that we are the only life in the universe: billions of chances does not matter if it takes trillions of tries. We could be here only if one of the tries worked, but it doesn't mean more than one must.

Reasoning from one example of life in the universe does not allow us to generalize to its distribution everywhere, so it remains impossible to predict the distribution of life in the universe. We can only optimistically (or pessimistically) hope it is out there and continue to look.

The fact that we do not know how likely life is should impact our reasoning about any observational evidence for alien life we might come across. As we discussed in the last chapter, biosignature candidates like atmospheric O_2, amino acids, and isotope fractionation are all subject to false positives—that is, they can be produced in the absence of life. These biosignatures, therefore, require us to know something about the prior likelihood of life to assess whether the observation could have been produced by aliens. If the origin of life is very common, then we might have high confidence the biosignature candidate was generated by an alien life-form. But the converse is also true: if we predict the prior probability of life to be low, such that the origin of life is very, very rare, we should not consider these biosignatures as evidence of biological activity, because the abiotic explanation is the more likely one, even if it is rare.

To understand how this works in practice, we can consider two potential claims of life detection, made for two different targets in the solar system, both published in the journal *Nature Astronomy* in 2021. The worlds in question were the planet Venus and the moon Enceladus. Venus is the second planet from the Sun, and is similar to Earth in its size, but not in terms of its surface conditions. Venus is the hottest planet in our solar system and has been long considered inhospitable to life. As such, it is mostly overlooked in astrobiology. Enceladus is an outer-solar-system moon orbiting Saturn, and it is only one seventh the diameter of our own Moon. It has been a target of interest to astrobiologists ever since the Cassini mission took photos of active plumes of water and organic material being spewed into space, providing evidence of a subsurface ocean of liquid water that could harbor life.

The claims of alien life that came in 2021 for these two worlds were based on observations of two different putative biosignature gases—phosphine on Venus, and methane on Enceladus.[20] One of these discoveries made international headlines, while the other was only read and discussed mostly by professional astrobiologists. Can you guess which is which?

Phosphine is among a suite of proposed candidate biosignature gases for exoplanet atmospheres. It is considered by some to be a good target in our search for alien life because it is a remotely detectable small molecule, and it is argued by some to be uniquely producible in select environments only by biological activity. Its detection in the upper atmosphere of Venus offered a chance to test ideas that researchers hoped could apply to exoplanets. There are two areas of debate with this detection: the first is whether phosphine is really in the atmosphere of Venus,[21] and the second is whether this should be interpreted as evidence of alien life in the clouds of Venus. The detection itself has been extensively debated—inferring the presence of specific molecules from spectral data is incredibly hard, even for our closest neighboring planets. But let's assume for a moment there is definitively phosphine in the Venusian atmosphere and focus on the claim that it is evidence of alien life in the cloud deck.

At the time of the public announcement of this discovery, there was so much excitement about the alien hypothesis that it even made the top science headline in *The New York Times*. However, phosphine—like most other biosignature gases—is subject to false positives because it can also be produced abiotically (and indeed we have detections of it in the atmosphere of Jupiter, where we do not assign it a biological origin). To disentangle whether the Venusian signal can be attributed to life requires calculating the likelihood it could be produced abiotically in the atmosphere of Venus. We cannot calculate the likelihood it is produced by life because we cannot guess what alien life on Venus would be like, including how it might metabolize to produce phosphine. The scientists reporting the discovery and its possible alien origin therefore ran a series of models to test the alternative hypothesis, seeking to determine whether they could explain the detection of phosphine abiotically. They concluded they could not.[22] Ruling out all possible abiotic mechanisms by exhaustive search is not in fact possible: we know very little about the chemistry and conditions of Venus, and certainly not enough to say we have constrained all possible geochemical explanations. Even so, the researchers ruled out some abiotic mechanisms, and concluded with a proposition that life might thereby be the most plausible explanation.

Evaluating whether alien life is the best explanation requires more than just ruling out any abiotic explanation. Additionally, we need to have some idea of the likelihood that life produced the signal: we cannot assign a likelihood for an abiotic production mechanism without knowing the corresponding likelihood of a biological one. Minimizing the false-positive rate of phosphine by attempting to exhaustively exclude abiotic mechanisms remains inconclusive if we do not know the prior likelihood for life to exist in the environment of Venus. If there is zero chance for the origin of life to have happened on Venus, and zero chance that life could evolve and thrive in the cloud decks, then it does not matter how low the probability of detecting phosphine is from abiotic production. The abiotic explanation would still be a better (higher probability) one! Without a hypothesis about how life could originate in such an environment, and what its metabolism could be, it is impossible to conclude a biological origin with the minimal data we have at hand. Significant claims about the source of phosphine (either biological or not) on Venus are premature: the hypothesis that phosphine is produced by life cannot be favored over the hypothesis of unknown atmospheric chemistry generating the signal because we don't have any way to predict the likelihood of life in that environment, or what it would be like. There is no explanation for life, and getting to an explanation should be our goal.

The Venus study based its evidence for possible life detection on the premise of ruling out abiotic explanations, which is a standard way of doing science in astrobiology right now in its preparadigmatic state. We lack consensus about what phenomena we are seeking to discover when we say we want to find to life, so we do the best we can with the tools we have.

The Venusian detection can be contrasted with what happened in the recent case of Enceladus, where the biosignature gas in question is not phosphine, but methane. Methane on Earth is produced via a process called methanogenesis, in which some types of microbial life produce methane as the product of their metabolic activity. Like phosphine, methane is a simple gas molecule: phosphine is a phosphorous attached to three hydrogens (PH_3), methane is a carbon attached to four hydrogens (CH_4), and both are low assembly index molecules with a molecular assembly of 1. Unlike with phosphine, however, metabolic pathways for producing methane gas are very well documented. Most of the methane that is in our atmosphere today is produced by biological activity. Methane has been proposed as a candidate biosignature gas for exoplanets that might be similar to the early Earth, which could have been covered in an orange haze. This vision of our early planet is sometimes called the "pale orange dot," a nod to Carl Sagan's "pale blue dot" description of modern Earth.[23] As with all other candidate atmospheric biosignature gases, methane is subject to false positive signals—it can, for example, be produced in volcanic outgassing and via serpentinization (a transformation of rock that releases energy).

Methane has been detected in the plumes of Enceladus, prompting some astrobiologists to wonder if it could be biological in origin. As with the case of phosphine on Venus, we want to confirm one of two hypotheses: (1) methane is produced abiotically, or (2) methane is produced biologically. We can compare these two scenarios by weighing their odds, that is, by comparing the likelihood of abiotic methane production to biological production based on known metabolic pathways from Earth's microbial life. If we do this, the likelihood of biological origin appears quite high, because what we know of the conditions on Enceladus could allow methanogenesis to work in this alien environment. Nonetheless, the team studying Enceladus's methane did not make international headlines. This is because their assessment of a biological origin for the methane did not just consider whether biology or abiotic activity could be the explanation; they also considered the prior likelihoods for these two explanations. If the prior likelihood is high for life to emerge in the ocean of Enceladus, then biological methanogenesis could be the best explanation. However, if we do not have a compelling theory for the origin and evolution of life on Enceladus, nothing can be definitively concluded, because the abiotic explanation is also possible.

The cases of phosphine on Venus and methane on Enceladus provide two examples of the challenges astrobiologists face in looking for life in the absence of a theory for what life is. The first is the challenge of false positives: with many biosignatures drawn from analogy to Earth's simplest biochemistry, there are abiotic processes that can produce the same signal. Therefore, we cannot be confident in life detection based on the signal alone, as this might lead to at best ambiguous and at worst false claims of the detection of alien life. This brings us to the second challenge, which is the problem of unconstrained priors: without any expectation about whether life should exist in a given environment, candidate biosignatures that can be produced abiotically cannot rule out the abiotic explanation.

These challenges increase if we want to look for life on exoplanets. Getting data from planets orbiting other stars is hard: astronomers are lucky if they can get just a few pixels indicating the presence of an exoplanet, and from this data they must also resolve features of these alien worlds, like their size and atmospheric composition, in order to infer whether life could be present. We've already discussed how O_2 is a favorite candidate biosignature for exoplanets. This is in part because O_2 is so easy to detect, even over astronomically large distances, relative to other atmospheric gases. On Earth, we have abundant O_2 because of the photosynthetic activity of organisms. But on an exoplanet, planetary scientists have modeled how O_2 can be produced on water-rich worlds because starlight can break apart water molecules (H_2O) into hydrogen and oxygen: the hydrogen escapes the planet, leaving behind an atmosphere rich in O_2 on a seemingly habitable world, because it also has water. This is just one scenario—imagine being the planetary modelers that have to exhaust all possible abiotic scenarios for producing an oxygen-rich atmosphere on a planet orbiting another star. As creative as these scientists are, it is too big an ask to exhaust them all. The unique challenges of exoplanet searches for life led two of my former PhD students, Cole Mathis and Harrison Smith, to pen a boundary-pushing manuscript declaring that the search for life on exoplanets is futile.[24] They were purposefully provocative in how they went against the grain of optimism for the relatively easy chemical correlates of life exoplanet scientists currently go after. Instead, they outlined just how hard confirming the alien hypothesis on a planet around another star will be—at least before we have a paradigm shift in how we do astrobiology. (It's a proud former-adviser moment when your past students have more provocative ideas than your own!)

There are two resolutions to the challenges of life detection I've posed. Either we need biosignatures that are not subject to false positives, or we need to determine the probability for an origin-of-life event. Assembly theory solves the problem of false positives, because a falsifiable hypothesis of the theory is that life is the only mechanism that can produce high assembly objects. Thus, as we continue to develop and validate assembly theory here on Earth, we can increase our chances that we will be able to detect alien life on another world and be confident in our detection. And indeed, we are currently working on getting assembly theory "flight-ready" by calculating how feasible it is to make a detection of high assembly molecules using current mass-spec instrumentation from prior and upcoming NASA missions. This could prove critically important for analyzing data from solar system worlds like Titan, which the NASA Dragonfly mission will visit in the coming decades. Titan is an odd world. It is a moon of Saturn that has liquid on its surface, and an atmosphere, made not of water but of hydrocarbons. The only water on the planet is ice—on Titan, water is the solid rock. Imagine sitting on an ice rock overlooking a hydrocarbon lake—an alien environment indeed. Agnostic detection methods like assembly theory will be critical for worlds like this because we cannot anticipate the chemistry of life—if life exists on Titan—to be anything like it is on Earth.

Even when using assembly theory to look for life elsewhere, we are still confronted with an urgency to resolve our expectations for how common life is. To do this, we need to solve the origin of life. Brandon Carter's arguments rested on us not knowing the probability that life will emerge. If we knew the mechanism we could constrain the odds. Frank Drake's equation is unconstrained because we do not know this probability, so we cannot estimate the distribution of life. We cannot rule out false positives, because we do not know if the alien hypothesis is ever a good one. We do not know how long or how many locations in the universe we will have to search because we do not know what the possibility is that we are not alone among the stars.

Many people think first contact with alien life will be life we encounter on another world. I disagree. We don't know if there is *any* life beyond our own Earth. The probability of life could be so low as to render discovery of alien life on another planet technologically intractable. Furthermore, aliens could be so different as to defy the expectations of what measurements we will make. By connecting the science of the origin of life and that of detecting alien life via a theory for life, we open another, faster route to first contact. Our best bet for making contact with an alien life-form in the near term may be to evolve it, from scratch, in the lab. While most people currently think alien life will be found out there in physical space, somewhere on a distant planet or moon, I am becoming increasingly convinced we are more likely to find it right here on Earth—but there is an even larger universe of chemical possibilities we may need to explore to find them.

Five

ORIGINS

It is fall in Tempe, Arizona. It is a time of year that people enjoy visiting, and a good time to assemble a small group at Arizona State University to push the frontier of how we think about origins-of-life research.

I am walking just about as fast as I can, but I can't keep up with Lee. I do not understand how he can walk so fast. It seems inhuman. I am, however, enjoying the cadence the click of my heels is making against the pavement as he walks and I jog and we talk: it is a happy sound and usually inspires productive thinking, especially in a caffeinated mind. I am talking almost as fast as Lee is walking.

We are discussing the origin of life. More specifically, we are discussing what size scale a successful experiment will need to be to demonstrate a *de novo* origin-of-life event. We want to discover alien life, not on another world, but in the lab. To do that, we require that whatever we make must be as decoupled from our example of life as possible. We need the experiment to include as little of our human and technological agency as we can, so we can understand how life could happen without us. We may need to search a very large volume of chemical space, as we do not know how easy or hard it is for our universe to make life from scratch. In fact, we want to use the experiment to constrain the cosmological probability for life to form: if we can constrain the probability and mechanisms of the physics that generate life, we will have contributed a fundamental new estimate of the likelihood of life, an estimate that is critical to guide our search for life elsewhere. Lee suggests we should think about this like a search algorithm, but instead of searching through the space of possible websites or computer programs we will be searching through the space of possible chemistries our universe can construct. We need an experiment that can simulate a planet, because nature's chemical search engine—in the absence of life and chemists like Lee—is planetary geochemistry.

The Super-Kamiokande experiment in Japan pops into my head. I wish I could say I have seen it in person. The visuals I have seen on my computer screen are stunning, and so is the inspiring idea behind the experiment. The iconic image is stacks of spheres of water, stories tall, surrounding a pool with a flotation device on which two researchers in hazmat-looking suits are seated. Super-Kamiokande is a high-energy particle physics experiment built explicitly to look for proton decay. The experiment must be extremely clean, because the event we are looking for is exceedingly rare. We have never observed a proton decaying. Not one in the entire universe. There are a lot of protons, too, so if they do decay it must be very rare indeed. In fact, Super-Kamiokande was designed to bound just how rare this process is. The longer we go without ever observing a proton-decay event, the longer the lifetime we know protons must have. We have spent millions of dollars looking for an event that may never occur in our universe—it might even be impossible. We are looking for it because some of our theories have predicted it might occur and because even *not* observing a proton decay will place hard constraints on what is possible in our universe.

Contrast this to the origin of life, an event we do know can happen, because it happened at least once in our universe. However, we do not know the likelihood that life could happen again because we do not know the physics of how it happens. We have not spent billions of dollars in a unified, concerted effort to solve this. The community of those aiming to solve the origin of life is young and lacking a compelling vision of how to proceed. On this Lee and I agree, which is why we started trying to build that vision together along with other friends and collaborators.

I am convinced enough to agree with Lee that the problem is akin to writing a search algorithm for matter, rather than computers. This makes it tractable for us to connect a theory for what life is to experiments that can test how it originates. We need to search chemical space to see if we can discover alien life—life with an origin as independent of our own as possible—so we can learn how it happens. If we can figure out how the universe can make life without our intervention, we will have cracked the problem.

As we walk, Lee is trying to calculate in his head how many robots running how many chemistry experiments will be needed. What volume of matter do you need to observe a spontaneous origin-of-life event? He is convinced, and convinces me, that the current technology in his lab can scale up. He thinks fast when he walks, but also notes I am struggling to keep up. We decide to get an Uber so we are not late. In fact, we cannot be late, because I organized the workshop we are headed to, together with Michael Lachmann, and knowing Michael, he'll be reluctant to start without us.

A white car pulls up. We get in and are greeted by a friendly elderly gentleman with a ponytail. Lee introduces himself as a chemist working at the University of Glasgow and explains how "we are trying to make new life." I can't help chiming in, "But not the easy way!"

That conversation marked the beginning of our moon-shot project to solve the origin of life and first contact with alien life at the same time. We are going to need a very large experiment for this one. Planets are not easy to simulate with physical material, but this is not a simulation we can run in a computer. We would not even know what to program into a simulation, because we don't know how chemistry generates life.

Likewise, we humans could not simulate the conditions just after the origin of our universe either, until we did the experiment to reveal the physics. To do that, we had to build large particle accelerators, including the Large Hadron Collider at CERN. The LHC accelerates elementary particles to very high energies and can recreate conditions that we think have not existed since the earliest moments of our universe. We needed to do this, and spent billions on the experiment, because without observing the physics in the lab we could not confirm our theories. Without doing the experiment we could not search for new physics that might exist only at those high energies.

We must build a physical experiment to simulate planetary conditions to search for the origin of alien life for a similar reason. We need to test our theories against reality, and we can do so only in experiments that can explore the high combinatorial diversity of the chemical universe. We need to do the experiment and not simulate it, because we need to have the opportunity to discover the surprises we cannot now anticipate along the way. The origin of life is yet to be discovered, and after nearly four billion years of evolution, our biosphere is now emerging a technosphere that may have sufficient intelligence to solve its own origin.

Lineages of Propagating Information

I am not my atoms, and you are not your molecules. We are part of the current instance in a several-billion-years-old lineage of propagating information that has structured matter on Earth since life first emerged on it. The first life never died. Not only did it not die, it bifurcated and generated all of us, from the diversity of trees to the diversity of minerals,* to my coffee cup and my keyboard on this computer I am typing on, and your book (or device) as you read (or listen to) this. Individuals, and indeed species, are temporary aggregates of information as it persists in our biosphere.

We are part of a lineage of events stemming from the origin of

*Many minerals exist in the absence of life, but not all. In fact, most of Earth's mineral diversity is caused by biological or technological activity, suggesting minerals are products of evolution and selection too. Minerals are a frontier for assembly theory because they are not finite, distinguishable objects, making it difficult to define copy number.

life: everything you do and everything you touch also becomes a part of that lineage. It is for this reason that our technology cannot be decoupled from the concept of life. And it is for this reason that solving the origin of life will be one of the most challenging technical feats our species may ever face.

We are taught to think about all the information in life as genomic. If we trace that history back in time, the multitudes of information wound through all known biological species alive now start to converge at a point in the distant past of our planet in what is often referred to as the Last Universal Common Ancestor, or LUCA for short. LUCA is termed "last common" because there may have been earlier examples of ancestors common to all life today; LUCA is just the most recent. It is unclear how close LUCA gets us to the origin of life on our planet, as we do not know how much evolution happened before LUCA. All the universal components we trace back to LUCA are encoded genomically by fiat (we infer LUCA from genomes), so it can feel like all the important information must be written in DNA. However, genes are not the only way information propagates in biology, and indeed certainly not in life more broadly.

Our biological lineages propagate information in the substrate of bioelectric fields, in the cytosol of the cell, and in cell membranes. We discussed nongenomic information in bioelectric patterning across tissues in planaria in chapter 2, when we touched on the work in Mike Levin's lab. An open question Mike's lab is working on is where the information that contains the shape of the full worm is stored. It seems this information is not local to the genome and is distributed throughout the entire worm in the pattern of bioelectric signaling, but we don't know exactly how this works. Two-tailed and two-headed worms provide evidence that this "nonlocal" information patterning must be there. Mike's two-headed worms can be split into two heads, and instead of growing new tails, each of the two heads will grow another head to regenerate a full-size two-headed worm. Two-headedness is heritable even though the genomes are no different than one-headed worms. The same is true for the no-headed worms, but they do not live as long because they have no motivation to eat, which doesn't matter because they have no way to eat anyway. Both the two-tailed and two-headed worms have the same genotype, which is also the same as the wild-type worms, so the information about morphology cannot be in the genome and must be stored elsewhere. This is what led to Mike's conjecture about bioelectricity playing a critical role in storing shape. Other examples of nongenomic information abound in biology. Another one Mike has pointed me to is how deer store memories of past trauma in their antlers. Antlers grow seasonally, so male deer will lose a set and grow a new one each year. If you make a nick in the antler in one season, that information is stored—nonlocally—and remembered by the tissue on the deer skull to regrow an antler with a deformation in the same position the following season.

We can expand the concept of the propagation of information in life even further than cellular lineages. Our intellectual lineages propagate via brains, paper, and computers. Take Albert Einstein. His intellectual lineage has had far more substantial reach than his biological one, impacting many of us alive today in one way or another.

Look around you. How much of your environment right now was built by something else alive? If you are indoors probably nearly all of it. If you are outdoors, probably also nearly all of it. Can you begin to imagine a lifeless world? Can you recognize all the "life" around you?

As you exist now, you contain a multitude of informational lineages that have aggregated to form you. Information needs to persist long enough to be copied for it to be information at all. In cells, this means that molecules need to survive degradation long enough to have their information copied, as is the case for DNA, or to be assembled repeatedly, as is the case for other biomolecules. Likewise for information propagating in our minds, if we could not convey it to one another, it would not continue to exist. Sometimes these two manners of propagating information are discussed in terms of genes and memes. These are two of the most obvious projections of the many ways information is structuring our reality. Like the genes propagating in your chemistry, the memes propagating in your mind are part of a lineage, and both are part of what we need to explain to understand life: as far and wide as we can imagine, *all* information on Earth has its origin at the origin of life.

This seems like a philosophical point, but it is critically important to designing origin-of-life experiments. Without an understanding of what life is, we have been putting it into our experiments without even recognizing it. We ourselves exist as the products of evolution, the current instances of a long lineage of information propagating on our planet. We cannot design an experiment that completely removes this history. Designing an experiment is a living process, as are selecting and purifying the reagents, controlling reaction conditions, and selecting a target to synthesize. An obvious source of information we put in experiments is the molecules we start with. A less obvious one is the target structures we aim to produce (e.g., we impose a goal on the experiment, which itself is an act of selection). Every experimental manipulation, even if we think we are modeling something abiotic, is itself an example of life, because we are selecting out of a huge space of possibilities to bring any object into existence.

How then can we design an experiment and expect to observe the spontaneous emergence of life, e.g., via the spontaneous emergence of design we did not put in? The boundaries of any experiment we make to produce an origin-of-life event will always be set by information and constraints that are themselves the product of an origin-of-life event. This problem is not widely discussed in the origin-of-life literature, but it should be. My collaborators and I are aiming to get ahead of it by identifying how to quantify our agency and the informational constraints we impose on chemistry when we do origin-of-life experiments. We need to do this if we are to solve how life happens without us.

To do so, we need to make an *information vacuum*, a term coined to describe the difficulty of what we must do by the AI scientist and podcaster Lex Fridman. Only if we can do that will we be able to determine how much assembly the universe is generating *de novo*, without life, without us, within the boundary of our experiment. The goal is to observe an alien origin-of-life event in the lab, with the recognition that it may be impossible to make truly alien life, because even our experiments are part of our lineage of causation in the universe. What we can do is make new life as alien as possible by removing as much information from our own lineage as possible and controlling for the rest. This is how we can crack the physics of life. This is how we can find out how the universe can make new life without life already being there.

The Origin-of-Life Scam

In 1973, internationally renowned chef Julia Child filmed a cooking session at the Smithsonian. But Julia was not cooking her famous French cuisine. Instead, she was cooking up an experiment in prebiotic chemistry. The recording brilliantly illustrates how origin-of-life experiments have been done in the twentieth century and in the twenty-first century so far. It provides a launching point for understanding the conceptual shift necessary to solve the origin of life in the lab.

In the video Julia is cooking a "primordial soup"—this is the actual technical name used in the scientific literature for the mixtures of prebiotic molecules that might have given rise to the first life on this planet. The idea of a primordial soup started with the work of Alexander I. Oparin and John B.S. Haldane in the 1920s, who independently proposed the idea that conditions on early Earth might have been suitable for supporting the synthesis of complex organic compounds. Simple organic compounds would have been concentrated in a soup-like mixture and undergone various reactions to produce more complex molecules, like polymers, and eventually life. Thus emerged the hypothesis that a primordial soup, cooked on the early Earth, could lead to the spontaneous formation of life.

In 1953, a graduate student named Stanley Miller did an experiment under the supervision of his PhD adviser, Harold Urey, to test the idea of the primordial soup.[1] The experiment he set up was a stunning achievement for the time. The premise was simple: mix basic chemical ingredients, including molecules believed to be present on the early Earth, and put these in a round-bottom flask exposed to what was believed at the time to be the composition of the early Earth's atmosphere, complete with electric shocks to simulate lightning. From this simple setup, aiming to simulate in a flask the prebiotic environment, a huge surprise emerged—amino acids found in living cells! This was viewed at the time as a stunning success, and some even thought it would be a short time before we would observe new life-forms crawling out of the flasks. That has not happened, of course. In fact, if you run the Miller–Urey experiment too long, and do not select a suitable time to extract identifiable molecules, you get a black tar of unidentifiable organic sludge.

Nonetheless, the Miller–Urey experiment launched the modern era of prebiotic chemistry, which constitutes the study of how the molecules life uses could be produced without biology. Indeed, in the Smithsonian video, Julia declares, "We're doing a recipe for the chemical building blocks of life," while she executes the "recipe" from Miller's experiment. She mixes up sodium chloride, sodium sulfate, potassium bromide, potassium chloride, calcium chloride, and magnesium chloride added to one liter of pure water, according to her recipe card. Just like a chef can follow a set of steps to make a delicious meal, the prebiotic chemist can follow a set of selected steps to make specific molecules.

Prebiotic chemists have gotten good at this and have cataloged in the literature recipes that allow the synthesis of all the major building blocks of life. Yet none of the recipes have created any life-forms.

This is the "origin-of-life scam," as Lee has provocatively called it. What Lee intends by this is not that origins-of-life researchers are scam artists—rather they deserve a huge amount of respect for working in a very challenging area of science. But, at the same time, it is easy to deceive ourselves we are on the right track because we are baking in the answers we expect to get. Currently we are putting far too much of the final product—the molecules selected in the evolution of life on Earth—into the design of experiments. Prebiotic chemistry, as currently conducted, is an attempted brute-force search to solve the origin of life. We can verify the result quickly because we know the answer prebiotic chemists are after—the molecules life on Earth uses. As an analogy, to understand why this might be an issue, we can equate this to classes for solving hard problems in computer science. Consider the molecules of life as prime numbers, and their directed synthesis in a prebiotic chemistry experiment the same as verifying that a number is prime. If you test a known prime to verify if it is prime, the calculation is easy to do. Hence, we can readily engineer experiments that produce molecules life uses because we are setting the target and engineering its design. However, identifying new primes *a priori* is not in general easy. In fact, it is in a completely different complexity class of computation. This is the essence of cryptography.

When life first emerged, the universe did not have a targeted synthesis for the molecules life would be made of. If it did, that would imply that the laws of physics had the design of life in mind. But that's not the reality: the universe made objects that had information about how to make other objects. The universe learned to make the life we have on Earth, and it is entirely possible that it learned to make life a different way in different places, and could learn to make new forms of life if we design the right experiment to test it. We need to focus not on what life is made of, but on how the universe acquires the information to make it.

The challenge with current prebiotic chemistry goes beyond the fact that we are engineering our synthesis with the targets of known life in mind. We are literally replacing the biological architecture of the cell with the technological architecture of the chemistry wet lab. We replace enzymes with round-bottom flasks, and metabolic pathways with a sequence of interventions by a chemist. All of this is directed at the synthesis of known biological molecules, outside of the context of the cell, which is deemed "prebiotic" when in reality it is technological. The situation is so confused, that some astrobiologists even label purified DNA, extracted from yeast, as abiotic for their life detection work![2] If we follow the lineage of information leading up to the generation of these molecules, it includes the long history of the evolution of life on Earth. Is there a way to quantify the information and historical contingency we put into the recipe? Is there a way to scale the problem so we can get actual (and hopefully new!) life-forms emerging and not just the simple molecular building blocks of life? Can we ensure that when a biosphere emerges a technosphere, the evolution of technology can indeed reveal the origin of it all?

Chemputing Life

The invention of computers in the last century revolutionized nearly every aspect of human life, allowing us to automate many things we do, making them more efficient and scalable. However, the computer revolution was not mirrored in how we use and interact with physical materials, because matter is not known yet to be universally programmable in the same way computers are. In particular, chemistry right now is still done much the same way as it was for centuries. If you were magically transported back to a chemistry lab from one hundred years ago, you might recognize much of the equipment. It's the same equipment you'll find in most modern chemistry labs.

To solve the origin of life, we need a revolution in the technology that allows us to do chemistry. This revolution should be comparable to the revolution in computational technologies of the last century. We need to build a search engine for matter that can allow us to explore chemical space in an efficient and scalable way.

To achieve the goal of building a chemical search engine, Lee invented a technology called the chemputer. Lee and his lab at the University of Glasgow developed the chemputer because they wanted a machine that could explore chemical space to search for new life-forms. But they raised the money to build it by pitching how they could 3-D print any molecule. That is, Lee envisaged the chemputer as a robotic device that could automate the process of chemical synthesis and produce any molecule on demand. As I write, chemists have to synthesize, by hand, pretty much every molecule we use for anything. We are not very far beyond what the alchemists were doing centuries ago—artisanal chemists are very talented at doing specialized reactions, but no individual chemist has the knowledge to do all of known reaction chemistry, and we need specially trained people to be in the loop to perform specific syntheses. Many reactions we can explore face very human limitations, like how they are almost universally based on what we have discovered so far (compounds, reaction mechanisms), require running on timescales a human can perform (e.g., when a human can be in the lab, and how long they can run an experiment, or how fast they can iterate), and they can be done only in sequence or in parallel on the timescale of a human mixing the proper sequencing. This limits our exploration of chemical space to only one or two steps beyond known reactions in any exploratory endeavor a lab chemist might do. This restricted exploration is apparent in prebiotic chemistry, too, which is why we have simple recipes like the one Julia Child followed that can make only simple components of life, like amino acids, and only under very tightly controlled reaction conditions. This way of doing things does not allow us to explore chemistry in a generalizable enough way to make molecules we don't know about, molecules that could be the foundations of discovering alien life-forms.

The chemputer is at the center of new efforts to digitize chemistry and make chemical space programmable in the same way that computers made machines programmable. It is the foundation of the technology we think will enable solving the origin of life. Indeed, Lee is CEO of a deep tech company building chemputers to make chemical space accessible. Just as SpaceX aims to get humans to Mars by making low-Earth orbit economically viable, Lee's company, Chemify, aims to build the technology that will solve the origin of life by making chemical space accessible, programmable, and affordable to explore.* When I was a cosmology PhD student at Dartmouth digging into the myriad definitions for the origin of life, I never thought the pursuit would involve going through Silicon Valley and becoming entrenched in the start-up world. But some problems are too big for any individual academic lab to solve, and indeed too big for academia alone to solve. We need new kinds of collaborations and technologies to do it. This is one reason I am advising the company—I want to see their technology succeed because I want to see the origin of life solved in my lifetime. We are not the only ones seeing the connection between new technologies and the solution to what life is; other companies are being started with similar goals in mind. My expectation is that some of the most disruptive technologies in the coming decades will be not in artificial intelligence, as many suspect, but in the exploration of embodied intelligence as enabled by advances in deep tech at the interface of biology and chemistry.

Many think that the innovation of the chemputer is the hardware—the robot that does the chemical synthesis. However, the real innovation is the universal programming language associated with it. To put this in context, we need a brief digression into the history of computing.

In 1936, Alan Turing published one of the most significant mathematical proofs in human history. Many regard it as the in-

---

*In full disclosure I am on the board of advisers of Chemify, but my role there is exactly as I write in this book: I am enthusiastic about making possible the technology—at scale—that will allow us to solve the origin of life.

vention of the modern computer. I first encountered Alan's work through his writings on morphogenesis, where he developed some of the first mathematical models that could explain the development of patterns in biology, like leopard spots, from the underlying chemical reaction kinetics. In artificial intelligence, he is known for his invention of the imitation game: the first well-defined test for humanlike intelligence in machines. In the imitation game, you have a conversation, and if you are sufficiently convinced that you are conversing with a human, then the machine has passed the Turing test.[3]

In his 1936 paper, Alan tried to determine whether there was a generic device or machine that was universal in such a way that it could compute any computable function.[4] Most simply described, a function takes an input and maps it to an output. An example is the function that adds 1 to a given input, which can be written as $f(x) = x + 1$. Algorithms are automatable procedures for computing the outputs of some functions. But algorithms are not the same as functions. Some functions cannot be computed by *any* algorithm. But for those functions that can be computed algorithmically, there are often many algorithms that can compute the same function. Thus we can separate all functions into two categories of interest: those that are computable, and those that are not.

Alan realized it is possible to formalize the concept of a machine that can compute any of the computable functions in a finite number of steps. That is, for *all* functions that can be calculated *exactly* algorithmically, there should exist a single kind of machine that can compute them all. These machines are now known as universal Turing machines. A Turing "machine" is not a machine in the sense that we are most familiar with: Turing's machines cannot wash your clothes, they cannot order food for you, and they cannot allow you to check your favorite social media app or call your mother. Although many, if not all, of those machines may in some part be based on Alan's ideas.

Instead, Alan's machine is a relatively simple mathematical object, consisting of an abstract device that can manipulate symbols on a tape according to a look-up table that instructs how to perform the computation. The memory in a Turing machine is modeled by an infinite tape marked out into squares. On each square a single symbol may be printed or erased, and symbols can be read, with operations performed on them sequentially by the machine. With this device, Turing demonstrated the possibility of a machine that can compute any computable function. It's not every function (it will never exactly output π), but still every computable function is a lot of functions.

Even restricting ourselves only to the space of things that are computable as Alan did, the universal machine we can envisage is not as powerful as it may seem at first. This is because even with the computational power given to a universal Turing machine by its definition, there is a limitation: you cannot in general prove whether a given algorithm will halt or not. That is, you cannot generically prove without actually running a given program on a computer whether the program will eventually terminate or instead run forever. Turing's discovery became known as the halting problem, and it has deep implications for our understanding of computation and the reach of what we can know. It indicates that even when you restrict yourself to a well-defined formal system, there are things that cannot be proven by an algorithmic procedure, no matter how clever the machine or agent trying to prove them, because to do so would require devising a description outside the system in question.

What Alan did is very abstract, but he was able establish the idea of universal computability and how any function that can be computed must be one we can break apart into steps an algorithm can use to transform the input into the output.

What does this have to do with chemistry and origin of life?

If we think about computation as a window into what is physically possible, Alan's insights raise the following question: Is it the case that the only physically possible objects are those that can be built by an algorithmic procedure? Can the real universe generate objects only algorithmically?

In assembly theory the answer is yes, because objects are defined as those that can be broken apart and built by a stepwise procedure. It might be this is the only way the universe can bring complex objects into existence: by evolving the algorithms to assemble them, where the information can be stored in discrete steps. If an object cannot be built from simpler ones by a set of discrete operations, it may be that it cannot exist at all—at least not in our universe.

John von Neumann was among the first to see the connection between the abstract idea of computable functions and the more physical idea of constructable objects. Like Erwin Schrödinger, Johnny did his work before the structure of DNA was understood. Among Johnny's achievements were the invention of game theory, the invention of the architecture of modern computers, and foundational contributions to quantum mechanics, to name just a few.

Johnny made an analogy between the idea of computation, which is abstract, and what he called construction, which he intended to describe physical operations, rather than computational ones.[5] Whereas computation takes a given input to produce a given output, construction takes a physical object as the input and produces another physical object as the output. If we take the example of incrementing by 1, as an abstraction this can describe a lot of things in the real world. One is marking the anniversary of your birthday. Another might be building a polymer like a protein by adding one monomer (amino acid) at a time. But you might also be building a tower of blocks and incrementing by one block to make the tower taller at each step. In all three cases the computation (the algorithm) is the same, but the physical implementation is very different.

Inspired by how a universal Turing machine can in principle compute any computable function, Johnny devised a maniac of an idea for another abstract machine he called the universal constructor. The universal constructor (UC) is an abstract device that can in principle assemble any constructable object. Here constructable is meant to imply how building that object is not forbidden by the laws of physics. It is an open question whether a true UC that can really build anything could actually itself exist. That is, we do not know if one is permitted by the laws that govern our universe or not. Proof that it is possible would require a proof that every possible object consistent with the laws of physics is constructable and that there can exist another object (or set of objects) to perform all possible construction tasks. David Deutsch, for example, regards whether a universal constructor is possible as one of the most fundamental questions in physics. If there is a physical analog of Turing's halting problem, this may not be possible to prove even in principle, because it would require proving the universe can output specific objects without first running the universe itself forward in time to produce them. This has implications for whether open-ended evolution is possible: if we cannot predict what will happen (what comes into existence) until it does, then this suggests evolution might truly be open. The closest approximation to a universal constructor that we know of in our universe—the one system we know that can build the most possible things—is our own technosphere. But our current technology cannot build everything, and it is unclear how close an approximation to universal any technology could get, or whether another kind of system that is even more universal might one day succeed it.

Like its sister abstraction the Turing machine, a universal constructor is intended to be an automated system. It requires a set of instructions—an algorithm—to perform a given task. A Turing machine, for all its computing power, does not know it is a Turing machine, and a universal constructor does not know it is a universal constructor either. Nor will these devices ever know what they are. Their job is to blindly compute or construct. To be able to compute in the case of a Turing machine or build in the case of a constructor, they need instructions. Thus, Johnny had his universal constructor read a tape, just as a universal Turing machine does.

As I mentioned, we do not know whether a universal constructor can exist that can build everything permitted by the laws of physics. It may be that Johnny's conceptual machine is merely an abstract idea and it can be made physical only as an approximation, just as with Alan's machine. But Alan's machine also inspired the development of real computers. And from Johnny's machine we can also finds parallels in real things that exist. I mentioned the technosphere already, but the go-to example for most is not technology but cellular life.

Within a cell the translation machinery is responsible for building proteins, and it can build, in principle, any protein composed of the twenty or so coded amino acids (with minor variations across the tree of life as to what constitutes a coded amino acid). Thus, it is "universal" over the set of all possible protein sequences. We can therefore roughly think of the "tape" as the genome (written in DNA or RNA) and the "machine" as the cellular architecture, including translation machinery, or proteins.* The space of buildable objects is then all protein sequences composed of the coded amino acids. Although this sounds like it is a small subspace of all possible biological objects (it does not include genomes, cells, tissues, societies, etc.), it is still an astronomically large space: making every possible 100-amino-acid protein with the 20 canonical amino acids would fill a volume equivalent to not just one universe, but a mole (10^{23}) of them. It is impossible in our universe to do this. Building a universal constructor would only imply that it is possible to make anything consistent with the laws of physics, but even with such a Godlike machine we could never make everything all at once: the space of everything that can exist is simply too vast for the resources of a universe that exists in finite time with a finite amount of matter to ever explore.

In fact, the cell is a poor approximation to a universal con-

*This is all a gross oversimplification of how biology works, but we are trying to draw out abstract ideas that can apply to very different scenarios, so such simplification is necessary.

structor. While it is universal over the space of all possible proteins, which is admittedly very large, it is not universal over all objects. It is not even universal over all possible chemical objects. A cell cannot make many molecules that human chemists can, like pharmaceutical drugs such as sildenafil citrate, and PG5, the largest stable synthetic molecule ever made, with about 20 million atoms.

Is it possible to make a universal constructor that is universal over all possible chemical objects? Or at least the subset of all possible chemical objects chemists can make in the lab? This is the question Lee asked, and to address it he built a universal programming language for chemistry. He took all the things human chemists use: standard flasks, sets of reagents, reaction conditions, etc., and built them into a programming language to automate all of known chemistry.[6] In short, the chemputer is a universal constructor for all chemical space we have explored so far in the chemistry lab. This means we can remove chemists from being directly involved in dangerous reactions and instead allow them to take on less manual labor and more creative work, like dreaming up new molecules and synthesis routes, because we have a machine that can do the hard work of building chemical objects by automating their synthesis. Once you automate known chemistry, you can program into the system novel synthesis to discover new chemistry, even following multistep synthesis routes inaccessible to the human chemist—hence its utility for drug discovery. We can ask for the first time, as Lee likes to, do chemputers dream of electric drugs?

But better yet, with the most pressing challenges in automating synthesis solved, we can scale it up to run experimental programs that would otherwise be impossible. It is with this scaling of the technology that we might solve the origin of life, not in thousands of isolated experiments distributed across the surface of our planet run in different chemistry labs by artisanal chemists (as in current prebiotic chemistry experiments) where there are no standards for comparing one experiment to another, but in a single set of coupled experiments run by robots piloted by artificial intelligence with standard protocols for processing data and declaring success.

Lee right now has several chemputers running in his lab with primordial-soup chemistries in them, which he is evolving by programmable iteration over possible environments—from order of mineral additions to pH level to what compounds are introduced when.[7] These are boundary conditions—information input by a robot into the experiment. But there are several things that separate this effort from previous ones. Because the chemputer has digitized chemistry—it runs explicitly on an algorithmic procedure—we can precisely quantify how much information we are putting into the experiment. Because we have assembly theory, we can quantify the assembly of the starting molecules and track how assembled things become over time to look for evidence of the emergence of evolution and selection within the boundary conditions of the experiment. These two things when combined allow us to quantify how much of an "information vacuum" we have created via the experiment itself being a part of a lineage of life, versus how much assembly the chemistry in the experiment generates *de novo*. Because the chemputer is automated, we hope to scale this up to search large volumes of chemical space all at once. Even if the origin of life is a rare event in chemistry, we still have a chance of finding it. We can possibly bound the probability for the origin of life even if we don't observe it, just the way Super-Kamiokande is bounding the lifetime of the proton by not observing events. Lee estimates we will need a modular array—a cloud of potentially millions of chemical reactors. This is how we will simulate enough planetary chemistry to bring aliens to life on Earth.

Animating Matter

"The problem is as soon as we make them, they start dying." This is the dilemma Abhishek Sharma, a team lead in Lee's lab, is facing when I visit the lab in Glasgow on a gloomy November day. We're watching some oil droplets scurry about in a dish. I had gotten to observe their births as Abhishek directed a visiting student to pipette a few drops of oil into the petri dish. As each drop descended onto the dish it quickly began darting about this way or that. Abhishek can coerce the droplets to display all sorts of complex behaviors by adding acid or base to the petri dish at specific points around its rim and by changing the mixture of chemicals that make up the droplet. In fact, he doesn't do this himself, because the entire system is automated and run by artificial intelligence and algorithms written by him, Lee, and others in Lee's lab.

The oil droplets are one of the reasons I was really excited to start collaborating with the team in Glasgow. Even looking at them in the petri dish, they look alive. Of course, they are not alive (not yet), but the thing we're trying to figure out is when might they be. We want to know if they can cross the threshold to being alive, and how will we tell when that happens. This is a rather complex problem because the oil droplets can display a lot of "lifelike" behaviors. Take, for example, oil droplet formulations that flock like birds, others that chase one another around the petri dish, and still others that appear to swim like jellyfish or undergo division.[8] Each of these behaviors can be selected in the oil droplets because they have a "genotype" associated to them that's controlled by the machine. But their genotype is not encoded in DNA; it is encoded in a computer. The droplets are composed of a ratio of four chemical compounds, and a machine-learning algorithm has been programmed to track interesting behaviors. It is the closed loop, from behavior to selection to new behavior, that makes this machine a selection engine. Behaviors can be selected algorithmically, and the formulation of the droplet can be slightly varied just like a real genotype: the new, mutated genotype is then constituted in a new petri dish and its behaviors observed. By using selection to program behavior into matter this way, Lee's lab has shown that genotype-phenotype maps can exist outside of genetic systems.

This is my first time seeing the oil droplets in person. For years, I had watched them only in videos. Watching those videos gave me the idea of trying to program real living behaviors into the oil droplets. My thought was that because an artificial intelligence is already monitoring their behaviors, perhaps we could take real videos of living organisms—say cells moving across a petri dish or even a real flock of birds or schooling fish or a jellyfish or tadpoles—and record their behavior. We could then train the software on the biological data, so it will impose selectivity on the droplet experiments matching as closely as possible the recorded behavior of real biological systems. In this way my naïve thought was we could animate matter by programming via a selection algorithm real living behaviors into the inanimate oil droplets. This led me down a series of rabbit-hole thought experiments to determine if we can make the droplets alive: if we can get the oil droplets to be indistinguishable in terms of their trajectory of motion and their behaviors from real living systems, are they then alive? In some sense this was a reimagining of Turing's famous test for intelligence but was instead asking whether we could tell if a lump of matter possesses any agency.

Like all questions in science—especially the deep ones—this is more complicated than it would seem at first. Our expectation is that there should be a real measurable difference between living behaviors and nonliving behaviors (if life is indeed a natural category). If we are serious about using the oil droplets as a system that could probe the boundary between nonlife and life, we should be able to quantify that difference. We should be able to observe the transition to life and identify when it is that the oil droplet becomes animated in the way we associate with living matter.

As a theorist it's easy to have ideas, but I was not prepared for the experience of seeing the experimental setup in person. In the videos online you see only the oil droplet, so it's easy to be deceived and think that is what is coming to "life" in the experiment. That is not the full picture. The droplet factory, as Lee's lab calls the setup,[9] is a cyborg of sorts. The oil droplets are in an arena with four sets of two injection sites each, which have a remotely controlled acid and base pipette system. A camera monitors the oil droplet behavior from overhead: this is read into a computer that controls the chemistry to decide the next steps of the experiment. There is a separate mount for three-dimensional imaging of the oil droplets to study their morphology and how the morphology might influence their behaviors. There's a rig that allows studying in parallel many kinds of droplets all at once. This seeming elaborate setup reminds me of the particle physics experiments I worked on as an undergrad. But here instead of probing the highest energy scales in the universe, we are trying to pull dead matter into the living universe, and I mean that in a very visceral way. This is what it looks like when a planet evolves sufficient technology to reach into the nonliving, low assembly universe and pull inanimate objects into the domain of living physics.

In assembly theory, one of the conjectures we have about building a theory of life is about the space of possibilities. Life exists in the space where there are so many possibilities, that to explain why some things exist and not others, we need to invoke new physics—e.g., the physics of assembly. I discussed this in terms of chemical space: the number of chemical possibilities is so large that we should never expect to exhaust them all, and we need new explanations to describe what we do observe. I think the same logic applies no matter what the space of possibilities is. Using the same logic we applied for molecular assembly theory, I think we can also apply the principles of assembly theory to trajectories to test the hypothesis that some behaviors are too complex and require too much memory to ever be observed, unless they are produced by a physical system that is itself deep in time because it is the product of evolution.

My idea of copying living behaviors to inanimate objects is a bit flashy and fun but it's not enough to reveal the physics that we need to understand. In the new physics I am outlining in this book, it is clear we can imprint things from life on nonliving matter and they will become "life" (in that features of them will be evolutionarily contingent). But this does not mean that we've made them alive in the sense that they are now creative in the universe.

Lee and I have a project right now that we are running with the oil droplets. Our goal is to try to evolve them, select on their behaviors, and allow them to build up enough memory to exhibit genuine goal-directed behavior. Goal-directed behavior and agency are both widely regarded as key features of life. Agency roughly corresponds to cases where we think a system appears to have its own causal control over what happens and is not subject to the whims of the environment around it. Goals allow agents to steer and choose among possible futures. Humans can be said to exhibit goal-directed behavior because we imagine possible futures and select among them in terms of how we act: that is, we must have a representation in our minds of what could happen to establish a goal. This suggests that the possibility space itself plays a role in goal-directed behavior because it is the possibility space that is selected on to construct the future.

There is a lot of interest in artificial intelligence right now, in understanding whether we can detect agency and goals in algorithms. I want to know if we can recognize them in matter first (and then we can apply that understanding to AI, which also has physical "bodies," albeit very different than biological ones). The distinction is not as stark between detecting life in matter and in algorithms as one might first think. They are both products of evolution. However, in chemistry it will be easier to prove there is new physics, and in turn understand that physics, because this is where the informational properties—the ones that are embedded in time and history—first appear, and do so most starkly.

In assembly theory we see how objects created by life are extended in time. It is this feature that suggests where goals can become physical. Objects that are deeper in time also have larger possible futures, and therefore more futures to select among. It's easy to see how this can be the case if you think about how the future is built from the past: if your past has more possibilities (large assembly), then your future (which is built combinatorially from the past) will have more possibilities too. Thus, as selection makes memory physical in material objects, it also makes the future more open. It allows histories of objects to recombine to construct new possibilities in what futures might unfold.

To do experiments you have to be able to measure and test things. Lee and I came up with an idea we now call a Turing test for goal-directed behavior. It's inspired by Alan Turing's famous imitation game, which we've encountered already. The main question Turing asked was whether it was possible for machines to think like humans can. If they can think, how could we tell? If a human interrogator cannot tell a machine is not human, the machine will have passed the test. Nowadays it's relatively easy for large language models to at least superficially pass a Turing test. But even in Alan's day it was unclear if this was really the right test for intelligence in machines. Many scientists are interested in the phenomena of consciousness—what it is to be an experiencing agent—and whether that is possible in a machine. But Alan was pragmatic, and he wanted a way to test for humanlike properties that was concrete and measurable. Because he made the test intentionally very simple, some have confused his heuristic with a definitive test for intelligence.

As I mentioned in our discussion of consciousness, I am interested in how it is that information, or more generally "abstract" things, like thoughts, words, and the patterns in genomes, can influence the physical world. That is, how can we think about these abstract things themselves as physical objects? I conjectured that consciousness may only be measurable as a collective property and that observing counterfactuals become actual could play a key role in our ability to measure it. More broadly, what I'm interested in is whether it is possible to measure something internal like representation by measuring external behavior. It would be good if we could devise a test that addresses the hard questions of life and consciousness, and thereby gets at exactly the features that Alan had to avoid formalizing in his test for machine intelligence. To do so may require something as radical as taking time seriously as a material property. By focusing on how agency requires time (only objects that are products of selection can have it) we might be able to relate the degree of agency to the size of possible histories and futures encoded in objects as temporally extended materials. Perhaps we could use the framework of assembly theory, which treats objects as causal histories with different depths depending on how much information is necessary to assemble them, to identify those that have true agency, i.e., those that are alive, as distinct from inanimate matter that might fake it. This should also be true for algorithms that can fake being alive.

The test Lee and I came up with is conceptually simple but technically will be very hard to implement. Take a physical object and present it with a "decision" to be made. The decision is set such that the different possibilities are physically equivalent. Let's say, for simplicity, that there are two possible outcomes of the decision: move left or move right. Assume there are no physical differences: no magnetic field differences, no energy differences, no gravitational differences, no difference in the geometry of the decision, etc., between left and right. This means at the moment in time of the decision, move left or move right are physically *indistinguishable* by any current law of physics. Let's say we have an oil droplet, and when presented with such a left or right decision it always goes to the left. This could occur only if left is in fact favorable for the persistence of the object in the future, because there is no instantaneous difference. To make this decision reliably, the object must have been trained through its history to react to informational cues in the environment that have no bearing on its immediate state. This relies on a symmetry breaking in the present (between left and right, which are equal) by a possibility in the future, which is built into memory in the object by its history. While this sounds like retrocausality it is not: we are not suggesting that the future affects the present. Instead, we are suggesting that the present contains objects with different recursively constructed depths in time, and the futures available to those objects depend on those depths. If an object includes a history that allows it to construct a future it persists in, then it can pass the test. If it cannot do this, the object will display no behaviors that cannot already be accounted for in current physics. Here the memory *is* the object. For example, the memory in DNA, in assembly theory, is all the ways of making a DNA molecule by recursive operations, whereas in a more standard paradigm it would be the meaning of the sequence of As, Gs, Cs, and Ts in the molecule. The former can be read out from the object anywhere; the latter requires a machine (a human or cell) to interpret the information within the context of the cell.

Back to the oil droplets. How can they be made to pass this test for agency? The answer is selection. Right now, the oil droplets do not survive very long. They start dying as soon as they are born. More likely right now is that they were never alive and that I am anthropomorphizing by assuming they die. But Lee and Abhishek hope to push them to longer persistence times, so their behaviors not only include evolutionary memory, but also possibly behavioral memory. Because the droplets are out of equilibrium, they decay. To increase their persistence time requires establishing fueling stations, or food stations, where the droplets can acquire more chemicals to sustain themselves. This is just like how you must eat so that you may keep persisting in time. The experimental design that Abhishek was showing me aims to eventually evolve the oil droplets and select on behaviors that allow them to persist—for example, by selecting for those that visit fueling stations to refuel. Selection will then operate over two timescales—an evolutionary timescale of mutations in the formulation, and a developmental timescale of learning by the oil droplet in the arena. Those that learn to persist will be those that might stand a chance of passing the agency test.

To model the future, you must have a past. This means that selection had to have occurred. You cannot have goals unless you have a system that was selected. In fact, if assembly theory is on the right track the mechanisms of selection and goal-directed behavior are one and the same: the only difference is whether we are looking at the mechanism backward in time (selection) or forward (goals). To make goal-directed matter you therefore need selection. Thus, the solution for animating matter is simple: we need only select oil droplets to persist long enough in time to evolve goals that allow us not just to select on the droplets, but for the droplets to select on their own futures within the constraints of their environment. Or at least we hope it's that easy. It's no minor technological feat to build a machine that can pull oil droplets across the inanimate/animate boundary. It's probably the case that the first life we bring into existence will barely be alive and quickly fade back into nonexistence, but it just might be the case that this is how we learn to animate nonliving matter.

Towards a Genesis Engine

You might worry that there's no way an oil droplet could display complex-enough chemistry to store its own goals. You might be right. When I introduced the chemputer I also introduced the idea of doing origin-of-life experiments at scale. In fact, the real goal of the experimental program I am introducing here, being developed in Lee's lab, is ultimately to integrate the automated droplet factor and the search of chemical space into one experimental machine, which would allow us to explore the huge possibility space of chemistry that might become animated. How many possibilities do we have to run before we find something that's alive? This is something we don't know the answer to yet. But we can work out whether we can get the origin-of-life moon shot off the ground—we need to build the software, chemputers, and lab equipment to search a large-enough volume of chemical space to discover the origin of life, or at least constrain the physics of how it happens. We need a team of the right minds in the room doing theory and experiments. My dream is that the origin-of-life science we do in the future may not be that different from searching the universe for new physics, which we currently do in massive international collaborations in particle physics and cosmology. If the physics underlying life is just as fundamental as that underlying gravity or quantum physics, we should be building a concerted international effort to solve it. It's not impossible—we know it's already happened once. A living cell is "just" a bag of chemicals that is animated by the memory it stores and the goals it acts on. Life is how the past connects to the future. The cells we find on this planet now are very deep in time indeed, birthed from the genesis engine that is our planet. As we transition to new stages of technology, we are on the precipice of building genesis engines that might allow us to start new lineages of life in the universe. We might just discover first contact was always going to be the creation of alien life right here on Earth.

Six

PLANETARY FUTURES

"So, where is everybody?" If you had been sitting in the cafeteria at Los Alamos National Labs in the summer of 1950, you might have overheard physicist Enrico Fermi blurt this question out in debate with several of his colleagues over lunch.* Enrico was arguing that, if there are other intelligences out there in the universe, we should have encountered them by now. The absence of audible or visible chatter from aliens when we look to the skies—what is sometimes called "the Great Silence"—is telling us something. The absence of evidence can itself be evidence of what we have yet to understand.[1]

Enrico's simple question became legendary in the research area popularly known as the search for extraterrestrial intelligence (SETI), which we encountered briefly in chapter 4. This is because

---

*His three colleagues in conversation, Edward Teller, Emil Konopinski, and Herbert York, all report slightly different variations of this same question.

it gives a framing to the most perplexing feature enveloping nearly all modern scientific discussions about alien life: *we have not observed any aliens*. If you are the first to popularly ask a pointed question, no matter how simple, you can sometimes get your name on it. In this case, the paradox of why we do not observe intelligent aliens even though you might want to assume they're out there is now referred to as Fermi's paradox.

A simple answer is there are no aliens and we are alone in the universe. This would be hard to prove in the absence of knowing what it is we are confirming the nonexistence of. Another answer is the aliens are out there but we have not observed them, at which point the challenge is to identify why we haven't.

Among the most widely cited resolutions of the Fermi paradox was that first proposed by the economist Robin Hanson. Robin's proposal is epistemologically bleak. He assumes the reason we do not see aliens is because they are either all dead or they were never born in the first place. He argues the Fermi paradox can be readily resolved if every example of life that evolves will eventually hit a filtering event. This "Great Filter"[2] is an unspecified event. The only important feature is that the vast majority of life in the universe will not survive it. The event could be a mass extinction driven by normal planetary processes on inhabited planets. We know of five major extinction events in our own planet's history, which have been driven by planetary glaciation, volcanism, and large impacts from extraterrestrial rocks. The most devastating of these was the Permian–Triassic extinction, which obliterated 96 percent of the species alive at that time. These did not wipe out life entirely, but in many cases were so utterly decimating they might have. It is not hard to imagine that a mass extinction event could wipe out all life on a planet. There is also the possibility that intelligent species could wipe themselves out. The last century saw the very real possibility of our own self-annihilation via nuclear holocaust. Many think we face yet another existential threat in this century due to climate change or due to the pace of advances in artificial intelligence. Furthermore, given that we know of only one example of life in the universe right now—us—it could be that the origin of life itself is the filter. That is, it is not that all life dies out at a certain stage but that it is never born.

While there are many examples we might draw on from our own history to support Robin's thesis, his argument does not depend on the specifics of what the Great Filter is. It matters only that life very rarely emerges, or that all life dies out eventually. Thus, just as individual organisms have a life-span, Robin proposes there is a life-span for biospheres or, if they have enough time in an average life-span to evolve further, technospheres.

As with most theories about extraterrestrial life, the Great Filter is more useful as a concept for looking inward, rather than outward. Indeed, most philosophical discussion of the Great Filter is focused on our own biosphere's mortality and whether we should expect life on this plant to thrive long into the future or if we are in the twilight of our planet's inhabitance. If the Great Filter happens early, say before the origin of life, and most planets never evolve life at all, then the odds are in our favor: our own biosphere is likely in the clear because we already passed the filter. But this also means we are probably utterly alone, with no other life to share our successes with. If the filter is instead something that happens in our future, we should be worried. An existential threat that kills off the vast majority of technological life that emerges in the universe still looms in our future.* We might not be alone at our level of intelligence, but we also may not last long enough to discover whether there are others.

The Great Perceptual Filter

SETI theories like the Great Filter have it the wrong way around. If we are using ideas that purport to look outward for others "like us" but really instead look inward, we may need to rethink our strategy. We may instead need to focus on deliberately looking inward to understand more about ourselves by trying to understand what we might look like from an outward perspective. That is, if we want to recognize other examples of life, maybe we need to ask what we look like from the outside.

The word "filter" probably makes you think of something like a sieve, which removes solid particles from a liquid or gas passed through it. This is in line with Robin's intended meaning: he envisioned the universe literally removing any intelligent species that evolve, or more dramatically, removing the possibility of them before they happen. But filters are also used on cameras and audio devices to filter light and sound, muting some frequencies out entirely or at least distorting them. Perhaps it is in this sense that we should be viewing the Great Filter, if there is one.

We can speak now about these different meanings of the word

*Note, the existential threat of artificial general intelligence, much hyped in recent times, is not an existential threat with respect to the Great Filter because the artificial intelligences would survive the filtering event even though we do not.

"filter" in our technologies because, over billions of years, the innovations of sight and hearing, among many others, emerged through evolution and selection. We do not know exactly what our planet looked like when life first emerged. In fact, neither did the life that existed at that time. Nothing alive then could see. The evolution of photon receptors and eventually eyes came later: these evolutionary innovations relied on many other innovations previously made over geological timescales by single-cell organisms. Multicellular animals like mammals, which utilize about seventy different specialized cells to see, further advanced the technology of sight, but only by building on what came before. The mantis shrimp evolved perhaps the most complex multicellular eye known: it has compound eyes that move independently and have twelve to sixteen pigments (our own human eyes have three).

More complex ways of "seeing" were later evolved by human societies when we invented microscopes and telescopes. Microscopes allowed us to see a previously unknown microscopic world, yielding discoveries ranging from biological cells to subatomic particles. Telescopes allowed us to see distant reaches of the universe, enabling the discovery of moons around other planets, planets around other stars, and even allowing us to see in the distant reaches of our visible universe the earliest galaxies and exotic and previously unimagined phenomenon such as pulsars. We could not "see" gravitational waves for the first 3.8 billion years of our history, but then we developed technology that allowed us to do so. These kinds of advances in perception are what the historian of science Claire Webb, director of the Future Humans program at the Berggruen Institute, calls "technologies of perception," that is, technologies that expand our ways of perceiving and understanding the world.

It is in this sense that our technology is a perceptual filter through which we evolve to see reality in new ways. The standard definition of technology is the application of scientific knowledge for practical use. Historically, where philosophy and technology intersect, the goal has been to apply old philosophical ideas to understand new technology. But, as David Chalmers has pointed out in his book *Reality+*, in the area of technophilosophy, this logic is inverted: technology can be used as a new lens with which to visit old questions in philosophy.[3]

In the spirit of SETI and anticipating the alien, we can also ask what new insights might be gained by taking a broader, non-humancentric view of what constitutes technology and how this can be used to reinvestigate old questions in philosophy and about life more generally. Technology relies on scientific knowledge, but scientific knowledge is itself information that emerged in our biosphere, as we have seen with the examples of satellites and technologically synthesized elements. If we were observing our own evolution from the outside, we might note that scientific knowledge enables things to be possible that would not be without it.

Technology, therefore, in the broadest possible sense, is the application of knowledge (information selected over time) that allows things to be possible that are not possible in the absence of that knowledge. In effect, technologies emerge from what has been selected to exist. They are also what selects among possible futures—and builds them. An example is the possible technologies being developed that would yield robust carbon removal: if successful, these could change the future evolutionary trajectory not just of humans but of a huge diversity of species on Earth, and indeed of the entire biosphere.

We are accustomed to thinking about technology as uniquely human, but in the broader definition I propose, there are many examples across biology too. Far earlier than humans, it was the biosphere that invented most technologies on this planet. A key theme of this book, and indeed the history of science in general, is that of unifications—things that were once viewed as different are later understood as manifestations of the same phenomenon, dramatically increasing our explanatory power. Unifying matter and computation, as we do in assembly theory to explain life, also has the implication that we should unify biology and technology as manifestations of the same fundamental process: selection on what gets to exist. Just like the objects of life might include pencils and satellites, so, too, technology might include wings and DNA translation. Photosystems I and II—multiprotein complexes found in plants and other photosynthesizing creatures—harvest photons to use light energy to catalyze reactions. As evolutionary innovations, these technologies radically changed the climate of Earth in the great oxidation event, a period about 2.5 billion years ago when cyanobacteria produced a large amount of atmospheric oxygen, which contributed to later conditions that allowed the evolutionary emergence of multicellular life.

A focus on Claire's idea of technologies of perception, and how these evolve over time, forces a deep connection between our anticipated planetary futures and our primordial past. In order for our biosphere to evolve telescopes, microscopes, interferometers, and atom smashers we had to first evolve photon receptors, and then eyes, among myriad other biological technologies. If in the past (say before the invention of the telescope and microscope) we were not able to see all of reality because our technology was not sufficiently advanced, what parts of reality can we not see now?

Is the Great Filter merely a feature of our own early developmental stage of technology? Could it be aliens are out there but we *literally* cannot see them? At least not yet?

If you are not sold on the idea that the Great Filter is perceptual, consider how much of life on Earth we would not know is here without technology. For most of human history, we were unaware of the legions of bacteria being born, living, and dying across the surface of everything in our environment—even within us. It took the technological innovation of the microscope in the seventeenth century for us to finally see a microscopic world teeming with life. This knowledge came with the work of two scientists, Robert Hooke and Antoni van Leeuwenhoek, both fellows of the Royal Society in England, who built simple microscopes capable of magnifying our eyesight by anywhere from 25-fold to 250-fold. The first indication we had of viruses was cryptic patterns in infectious diseases they cause, but their existence was confirmed only in the late nineteenth century, when we evolved filter technology with fine-enough pores to sort out bacteria but not the tiny viruses. We also did not know about the thriving ecosystems in the darkest depths of the ocean floor until the second half of the twentieth century, when submarines that could stand intense pressures got us close enough to observe them.

Every advance in how biology generally, or us specifically, has come to see and understand the world better has been accompanied by a technological advance.

Technology has enabled us to see life in new ways. But more remarkable is how at the same time technology is *life* (even as it may not yet be alive).

Technogenesis: A Second Origin

History teaches us that the discovery of new forms of life first requires the advent of technologies that allow us to sense and explore the world in new ways. But almost never do we consider those technologies themselves as life. A microbe is life, and surely a microscope is not. Right? But what is the difference between technology and life? Artificial intelligences like large language models, robots that look eerily human or act indistinguishably from animals, computers derived from biological parts—these are revealing a boundary between life and technology that is increasingly blurred. This blurriness should be familiar to us now in this book, as it is the same kind of blurred boundary we discussed in chapter 1 that separates chemistry from life, and indeed it is the same blurriness that frustrates our attempts to define a clear boundary to what life is.

Among the most widely discussed technologies emerging now on this planet are artificial intelligences. These are often discussed as disembodied, and worse, outside of any evolutionary context. But the technologies we are inventing now—from large language models, to computer vision, to automated devices—are not separate from the biosphere that generated them. Rather these represent the recapitulation of life's innovations into new substrates, and they are allowing the emergence of intelligent life at a new scale—the planetary.

A key feature of assembly theory is historical contingency: new objects come into existence only because there is a history that supports their formation. We see this same kind of contingency in all biological and technological innovations. Multicellular eyes could not evolve before cells with photon receptors any more than ChatGPT could evolve before human language—both rely on previous developments in a lineage of evolving technologies.

We are in the midst of another transition from the biological to technological: We are using algorithms to interpret data and "see" the world for us. To understand data from our telescopes we must write algorithms to process them. To interpret data from many of our microscopes—things like the Large Hadron Collider, which collects terabytes of data on particle interactions—we must do the same. This is not unlike how the brain had to co-evolve with the multicellular eye to process the frequency of light data that the eye receives in order for our mammalian brains to construct the images we "see." Brains and eyes evolved together to reconstruct in mental images some features of the external world.

Life is not just a property of individual organisms, but a feature of our planet as a whole, because all the lineages are connected across time and are globally distributed. The fundamental unit of life is the entire structure of information-patterning matter on this planet, with a temporal size of a few billion years and a spatial extent the size of the Earth. Viewing life at the planetary scale, we might expect to see the same features recurring across time at new levels of organization, as each evolved feature gradually scales up to the planetary. The Gaia theory, popularized in the work of James Lovelock and Lynn Margulis, was intended to conceptualize how life has established feedback loops with the planet that allow it to maintain itself over time.[4] In some sense, Gaia views the entire planet as a living entity because it is planetary-scale self-regulation of the biosphere that allows life to maintain itself over eons.

But Gaia theory does not address the hierarchy of complexity that life evolves over time—that is, the major transitions of life recurring across scales, from molecular to cellular, to multicellular to societal, to multisocietal to planetary. As the evolutionary biologists Eörs Szathmáry and John Maynard Smith have pointed out, there have been several major transitions in the evolutionary history of Earth associated with what counts as an evolutionary individual.[5] These include the transition from chemistry to cellular life, prokaryotes to eukaryotes, asexual clones to sexually reproducing populations, single-celled to multicellular organisms, solitary individuals to societies, and societies to language. Each of these major evolutionary transitions is associated with new modes of information processing and storage, and new collective modes of being. What they left off is the transition from societies to global culture: the multisocietal aggregates that have only very recently emerged on this planet are made possible through the interaction of linguistic human societies. Artificial intelligence is among the new modes of information processing and storage necessary for this new transition in the hierarchy of major evolutionary events. AI is a feature of the storage of information and its processing at the scale of human societies—a scale our brains are not equipped or evolved to process individually.

What is emerging now on Earth is planetary-scale, multisocietal life with human brain-like functionality. AI makes possible integrating data across the different technologies of perception we have evolved. It is hard for us to see this as life because it is ahead of us in assembly, not behind us. As I mentioned, this means these things look virtual rather than physical to us, because they are hyperobjects with a much larger size in time than we have, yet they are composed of us in the same way we are composed of individual cells and, below that, molecules. Furthermore, it is hard to fathom this because we are accustomed to viewing life only on the scale of a human life-span, not in terms of the trajectory of a planet.

The discovery of alien life can be built into the future trajectory of life on Earth, but we must first come to be able to see and to understand what we are, so we can recognize things like us that might be out there, or even those that could be here but we cannot imagine. We need to build technology sufficiently intelligent enough to see us from the outside, and theories to explain what we see. Assembly theory offers one way to do this, because it gives a window into how to see life as complex objects that build other objects, where the objects themselves are distributed in time. The artificial intelligences we are creating are another key piece because they will allow us to process data at a large-enough scale to see objects that are as deep in time as we are. In this sense artificial intelligences are also a technology of perception: allowing us to observe ourselves in a fundamental, more abstracted, way for the first time by taking the vast volumes of data that describe us as a collective (as societies) and compressing these data down to lengths and timescales that individual humans can sense and interact with. ChatGPT can be alienating to some, not because it is a human-level intelligence (as is too much discussed) but because it compresses societal-level intelligence down to a human scale (one can argue whether "intelligence" is the right word here).

The emergence of a technosphere may be precisely what is required for a biosphere to solve its own origins and therefore to discover others like it. To make this transition and make first contact, it may be critical to where we sit now in time that we recognize how thinking technologies are the next major transition in the planetary evolution of life on Earth. It is what we might expect as societies scale up and become more complex, just as life simpler than us has done in the past. The functional capabilities of a society have their deepest roots in ancient life, a lineage of information that propagates through physical materials. Just as a cell might evolve along a specific lineage into a multicellular structure (something that's not inevitable but has happened independently on Earth at least twenty-five times), the emergence of artificial intelligences and planetary-scale data and computation can be seen as an evolutionary progression—a biosphere becoming a technosphere. The question now is at what point does a biosphere emerge sufficient technology to solve its own origins, and how does this transition the future evolution, when planetary intelligence[6] is achieved?

In a conversation with Paul Davies about the struggles of working across so many fields of research, he pointed out that it is easy to be mistaken as contributing superficially by throwing out what seem on the surface like many random and disconnected ideas. What most people do not see, he continued, is how if you drill down deep enough in your thinking, different things can begin to look much the same. What we are after in this book—to understand what we are—may already be right in front of us. We cannot see ourselves clearly because we have not built a theory of physics yet that treats observers as inside the universe they are describing: that understanding is muddled across seemingly disparate concepts we refer to as "matter," "information," "causation," "computation," "complexity," and "life." Assembly theory is an attempt to see all of these as the same thing. Because these concepts seem so different to us right now, a theory at their intersection must necessarily hit very deep into the nature of how we describe our reality. It will not be immediately intuitive, just like the curvature of space-time around you right now is not at all intuitive but has explanatory power for how we think about and interact with the reality in which we live.

Lee and I recently visited Michael and Chris at the Santa Fe Institute (SFI) to continue our work on assembly theory, and in particular to lay the technical groundwork for proving why we should not expect any other planet to evolve life that is like our own. The four of us do not just want to solve how life arises in the universe, but also to anticipate life as no one yet knows it. On that visit, Lee and I also paid a visit to David Krakauer at the SFI's new Miller Campus to get advice on our ideas for an origin-of-life moon shot. We explained how we want to make aliens in the lab to prove the mechanism for the origin of life and, more importantly, prove the explanation for what life is. We did so amidst a rumbustious tour of construction sites: David is among the most visionary individuals I have yet met, and he sees a future environment for his institution that fosters the maximal creative potential of the collection of minds that come through it—inclusive of a parkour course, which he described with much animation. Sitting down to dig into our discussion more after the tour, David's response to the idea of an origin-of-life moon shot was a bit surprising. "We already know what life is!" he remarked. And indeed, he is right that many of us who think about the problem deeply already share a common sense of what life is—much work is converging on the idea that information must be at its core. But establishing a new paradigm takes massive effort. We can push to do more to understand the knowledge we are generating. Life exists in biology and technology as far as we know, but we do not understand the transition or continuation of life from biology into technology, any more than we understand the continuation of life when chemistry transitions to biology. Because of this we cannot yet anticipate the possibilities for what our own future holds.

Is life really *the* hard problem to be solved at this moment in our history? I am increasingly convinced it is, not just because the theory and experiments I describe in this book are advancing so rapidly, but because we are currently collectively so lost in understanding the technological transitions we are living through, as many objects in our environments are seemingly being animated before our very eyes. How are we to understand the artificial life and intelligences we create if we do not understand what life is? Presumably any planet with life hits a critical precipice where it must understand its own deepest evolutionary origins in order to understand what its future might hold and to steer it. Solving the origin of life, and indeed what life is, is therefore one of the most critical transitions that can happen in the evolution of any living world. I, for one, want to see it happen in the instance of my lineage where I am still alive.

ACKNOWLEDGMENTS

This book is already intended as an acknowledgement of the many intellectual influences I've had in my career. It is a true privilege that my trajectory has allowed me to engage with so many amazing minds. While I do mention some key conversations that have challenged and deepened my thinking, I must also note that many (if not most) of the important ones have been left out of the discourse of this book. It would simply be impossible to put the sum of all conversations and debates that have influenced me into a single volume.

There are a few individuals who have had an outsized impact on my thinking and my approach to deep scientific questions. And so, I want to take the time here to thank them specifically.

First and foremost is Lee Cronin, a collaborator and friend. Lee has a remarkable mind, and his thinking and his standards for scientific inquiry make him stand apart from any other individual I have met. Before I came to know Lee and his work, I was not optimistic that the problem of the origin of life would be solvable in my lifetime. Even if I could find a way through to the conceptual foundations that might solve the problem, I was not sure anyone would understand what I was trying to do. The experiments I was learning about in prebiotic chemistry seemed insufficient to solve the problem anytime soon. Lee convinced me otherwise, and working with him has given me tremendous hope that we can do it, and soon. Lee's ability to break through walls and build entirely new experimental and theoretical paradigms has been the greatest inspiration of my career. He also has an uncanny ability to push me exactly as hard as I want to push myself. Of Lee, I can say that knowing him has made me a better scientist and indeed a better human.

I also want to thank my mentors, who have encouraged me to work on deep problems and have taught me that guiding someone's development of their own distinct skills to ask and answer fundamental questions is just about the best gift you can give a young scientist. Marcelo Gleiser was an early influence in this regard, and without his encouragement I would never have started working on the origin of life in the first place. Paul Davies gave me a remarkable opportunity on day one of my postdoctoral appointment in the Beyond Center, when he tasked me with putting aside all the current dogmas and seeing if I could come up with what I thought the key unanswered problem of the origin of life was. This seemingly simple task has entirely shaped my career. In addition to being a constant source of challenging fundamental puzzles, Paul has also been just about the best champion I could have asked for as I transitioned through various career stages. I am grateful to him and to Pauline Davies for their friendship and unwavering support.

The flip side to being mentored is mentoring, and I have learned as much from my mentees over the years as I have from my mentors. I want to thank the many PhD students, postdoctoral scholars, and undergraduate students that I have been honored to have in my lab over the years. Sharing their passion for tackling tough questions and their willingness to push boundaries has been one of the greatest joys of my career. I could not have asked for a better team.

This book was mostly written over the height of the COVID-19 pandemic, and I met regularly on Friday afternoons with Lee, Christopher Kempes, and Michael Lachmann via Zoom during those days. So, in addition to Lee, I want to thank Chris and Michael, too, as I am grateful to the three of them for many, many deep, humorous, and just plain fun conversations. They were a wonderful diversion during a tough period in my life, as I juggled the tremendous demands of running a large lab remotely while simultaneously working on this book, and the ideas contained herein, and caring for small children at home (who have been utterly brilliant through it all).

It probably goes without saying, but should nonetheless be said, that I am grateful to my parents Lynda Pellegren and George Pellegren, who throughout my childhood and into my adult years instilled in me a nonconformist, creative mindset that they were happy to watch unfold as I grew into who I am today. (Despite having no idea that I would choose to express myself through the medium of physics!) I can be sure of who I am because they set the boundary conditions for me to find myself.

I also owe immense thanks to Courtney Young, my editor at Riverhead Books, who was fearless in suggesting edits as the manuscript progressed. I am grateful for her willingness to gently push me in directions that have dramatically improved the text. I am also thankful to my agent, Max Brockman, who early on embraced my vision for the book and supported me in making it happen. I must also thank Arizona State University and the many wonderful colleagues I have there. I fear if I had started my career in a different academic environment, my science would not have grown as it has. I was instead very lucky to land at ASU where the embrace of, and support for, innovative approaches and new ideas is palpable.

It has been a joy and a privilege to write this book, and I hope it has inspired you to think just a bit differently than you did before. I hope to continue to propagate the gift of inspiring deep thought and new thinking that others have given me.

NOTES

1: WHAT IS LIFE?

1. Jack W. Szostak, "Attempts to Define Life Do Not Help to Understand the Origin of Life," *Journal of Biomolecular Structure and Dynamics* 29, no. 4 (2012): 599–600.
2. Modern physics arguably may have started with Galileo Galilei, who invented many of the concepts that now form the foundations of modern science and physics in particular, especially in his efforts to use experimentation and observation to inform his understanding and explanations of reality.
3. Carlos Mariscal, "Life," Stanford Encyclopedia of Philosophy Archive, Winter 2021 Edition, https://plato.stanford.edu/archives/win2021/entries/life.
4. Gottfried Wilhelm Leibniz, "The Monadology, 1714," *Gottfried Wilhelm Leibniz: Philosophical Papers and Letters*, ed. and trans. Leroy E. Loemker (Dordrecht, The Netherlands: Kluwer Academic Publishers, 1969).
5. Mary Shelley, *Frankenstein: The 1818 Text* (New York: Penguin Classics, 2018).
6. Marco Piccolino, "Animal Electricity and the Birth of Electrophysiology: The Legacy of Luigi Galvani," *Brain Research Bulletin* 46, no. 5 (1998): 381–407.
7. Darwin wrote about spontaneous generation in his last poem, "The Temple of Nature," published in 1803. For more of his writings on life and its evolution, see Erasmus Darwin, *Zoonomia, or the Laws of Organic Life. In Three Parts* (Boston: Thomas & Andrews, 1809).

8. *Jacobellis v. Ohio*, 378 U.S. 184, 1964, https://supreme.justia.com/cases/federal/us/378/184.
9. Carol E. Cleland and Christopher F. Chyba, "Defining 'Life,'" *Origins of Life and Evolution of the Biosphere* 32 (2002): 387–93.
10. Carl Sagan, "Definitions of Life," in *The Nature of Life: Classical and Contemporary Perspectives from Philosophy and Science*, ed. Mark A. Bedau and Carol E. Cleland (Cambridge, UK: Cambridge University Press, 2010), 303.
11. David W. Deamer and Gail R. Fleischaker, *Origins of Life: The Central Concepts* (Burlington, MA: Jones and Bartlett Learning, 1994).
12. Walter Gilbert, "Origin of Life: The RNA World," *Nature* 319, no. 6055 (1986): 618.
13. Lucas John Mix, "Defending Definitions of Life," *Astrobiology* 15, no. 1 (2015): 15–19.
14. Erwin Schrödinger, *What Is Life? The Physical Aspect of the Living Cell. Based on Lectures Delivered Under the Auspices of the Institute at Trinity College, Dublin* (Cambridge, UK: Cambridge University Press, 1943).
15. David Deutsch, *The Beginning of Infinity: Explanations that Transform the World* (London: Penguin Books, 2011).
16. Léon Brillouin, "The Negentropy Principle of Information," *Journal of Applied Physics* 24, no. 9 (1953): 1152–63.
17. Here I am referring to a recording of the observation of the Great Red Spot in 1831, after which the spot became regularly recorded. There are possible earlier observations; the earliest is Robert Hooke's observation of a storm on Jupiter in 1664, but it is unclear if this observation was an earlier storm or is indeed the same storm that persists on the planet today.
18. Brian Greene, *Until the End of Time: Mind, Matter, and Our Search for Meaning in an Evolving Universe* (New York: Vintage Books, 2021).
19. Daniel C. Dennett, "Herding Cats and Free Will Inflation," Romanell Lecture, American Philosophical Association, Chicago, Illinois, February 28, 2020.
20. Philip W. Anderson, "More Is Different: Broken Symmetry and the Nature of the Hierarchical Structure of Science," *Science* 177, no. 4047 (1972): 393–96.
21. David Kushner, "Cormac McCarthy's Apocalypse," *Rolling Stone*, December 27, 2007, 1–8.
22. David Krakauer, "The Complexity of Life," SFI Bulletin, Spring 2014, https://sfi-edu.s3.amazonaws.com/sfi-edu/production/uploads/publication/2016/10/31/Bulletin_April_2014_FINAL_1.pdf.

23. Thomas S. Kuhn, *The Structure of Scientific Revolutions* (Chicago: University of Chicago Press, 1962).

2: HARD PROBLEMS

1. David J. Chalmers, "Facing Up to the Problem of Consciousness," *Journal of Consciousness Studies* 2, no. 3 (1995): 200–219.
2. Galen Strawson, "Consciousness Isn't a Mystery. It's Matter," *New York Times*, May 16, 2016, 16.
3. Hedda Hassel Mørch, "Is Matter Conscious?" *Nautilus*, March 31, 2017, https://nautil.us/is-matter-conscious-236546.
4. Annaka Harris, *Conscious: A Brief Guide to the Fundamental Mystery of the Mind* (New York: Harper, 2019).
5. Daniel C. Dennett, "Facing Up to the Hard Question of Consciousness," *Philosophical Transactions of the Royal Society B: Biological Sciences* 373, no. 1755 (2018): 20170342.
6. Albert Einstein, "What Life Means to Einstein: An Interview by George Sylvester Viereck," *Saturday Evening Post*, October 26, 1929.
7. Sara Imari Walker and Paul C. W. Davies, "The 'Hard Problem' of Life," in *From Matter to Life: Information and Causality*, ed. S. I. Walker, P. C. W. Davies, and G. F. R. Ellis (Cambridge, UK: Cambridge University Press, 2017), 19–37.
8. Carole Lartigue et al., "Genome Transplantation in Bacteria: Changing One Species to Another," *Science* 317, no. 5838 (2007): 632–38.
9. Fallon Durant et al., "Long-term, Stochastic Editing of Regenerative Anatomy via Targeting Endogenous Bioelectric Gradients," *Biophysical Journal* 112, no. 10 (2017): 2231–43.
10. Michael Levin and Christopher J. Martyniuk, "The Bioelectric Code: An Ancient Computational Medium for Dynamic Control of Growth and Form," *Biosystems* 164 (2018): 76–93.
11. David Grinspoon, *Earth in Human Hands: Shaping Our Planet's Future* (London: Hachette, 2016).
12. David Deutsch, "Constructor Theory," *Synthese* 190, no. 18 (2013): 4331–59.
13. Chiara Marletto, *The Science of Can and Can't: A Physicist's Journey Through the Land of Counterfactuals* (New York: Viking, 2022).
14. David Deutsch and Chiara Marletto, "Constructor Theory of Information," *Proceedings of the Royal Society A: Mathematical, Physical and Engineering Sciences* 471, no. 2174 (2015): 20140540.

15. P. C. W. Davies, "The Problem of What Exists" (2006), https://arxiv.org/abs/astro-ph/0602420.
16. Stephen Hawking, *A Brief History of Time* (New York: Bantam Books, 1988).
17. Max Tegmark, "The Mathematical Universe," *Foundations of Physics* 38 (2008): 101–150.
18. Regine S. Bohacek, Colin McMartin, and Wayne C. Guida, "The Art and Practice of Structure-Based Drug Design: A Molecular Modeling Perspective," *Medicinal Research Reviews* 16, no. 1 (1996): 3–50.
19. Stuart A. Kauffman, *At Home in the Universe: The Search for Laws of Self-Organization and Complexity* (New York: Oxford University Press, 1995).
20. Stuart A. Kauffman and Andrea Roli, "A Third Transition in Science?" *Interface Focus* 13, no. 3 (2023): 20220063.
21. Charles Darwin, *On the Origin of Species by Means of Natural Selection, or, The Preservation of Favoured Races in the Struggle for Life* (London: J. Murray, 1859).

INTERLUDE: NEW PHYSICS

1. Krzysztof Belczynski et al., "The First Gravitational Wave Source from the Isolated Evolution of Two Stars in the 40–100 Solar Mass Range," *Nature* 534, no. 7608 (2016): 512–15.

3: LIFE IS WHAT?

1. Sara Restuccia, Graham M. Gibson, Leroy Cronin, and Miles J. Padgett, "Measuring Optical Activity with Unpolarized Light: Ghost Polarimetry," *Physical Review A* 106, no. 6 (2022): 062601.
2. Stuart M. Marshall et al., "Formalising the Pathways to Life Using Assembly Spaces," *Entropy* 24, no. 7 (2022): 884.
3. Scott Aaronson, Sean M. Carroll, and Lauren Ouellette, "Quantifying the Rise and Fall of Complexity in Closed Systems: The Coffee Automaton," (2014), https://doi.org/10.48550/arXiv.1405.6903.
4. James P. Crutchfield, "The Calculi of Emergence: Computation, Dynamics and Induction," *Physica D: Nonlinear Phenomena* 75, nos. 1–3 (1994): 11–54.
5. Sean M. Carroll, "Why Boltzmann Brains Are Bad," in *Current Controversies in Philosophy of Science*, ed. S. Dasgupta, R. Dotan, and B. Weslake (New York: Routledge, 2020), 7–20.

6. Sara Imari Walker and Paul C. W. Davies, "The Algorithmic Origins of Life," *Journal of the Royal Society Interface* 10, no. 79 (2013): 20120869.
7. Stuart M. Marshall et al., "Identifying Molecules as Biosignatures with Assembly Theory and Mass Spectrometry," *Nature Communications* 12, no. 1 (2021): 3033.
8. Silke Asche et al., "What It Takes to Solve the Origin(s) of Life: An Integrated Review of Techniques" (2023), https://arxiv.org/abs/2308.11665.
9. This quote is taken from Sir Roger Penrose's interview on the *Lex Fridman Podcast.*
10. Frank Wilczek, "Physics in 100 Years," *Physics Today* 69, no. 4 (2016): 32–39.
11. Isaac Asimov and Jason A. Shulman, *Isaac Asimov's Book of Science and Nature Quotations* (London: Weidenfeld & Nicolson, 1988).
12. See, e.g., David Deutsch, *The Fabric of Reality* (New York: Penguin Books, 1998).
13. George Ellis and Barbara Drossel, "How Downwards Causation Occurs in Digital Computers," *Foundations of Physics* 49, no. 11 (2019): 1253–77.
14. Timothy Morton, *Hyperobjects: Philosophy and Ecology After the End of the World* (Minneapolis: University of Minnesota Press, 2013).
15. Chiaolong Hsiao, Srividya Mohan, Benson K. Kalahar, and Loren Dean Williams, "Peeling the Onion: Ribosomes Are Ancient Molecular Fossils," *Molecular Biology and Evolution* 26, no. 11 (2009): 2415–25.
16 David Deutsch, *The Beginning of Infinity: Explanations that Transform the World* (London: Penguin Books, 2011), 1.

4: ALIENS

1. David S. McKay et al., "Search for Past Life on Mars: Possible Relic Biogenic Activity in Martian Meteorite ALH84001," *Science* 273, no. 5277 (1996): 924–30.
2. For a detailed account of the sequence of events around the pulsar discovery, see, e.g., Alan John Penny, "The SETI Episode in the 1967 Discovery of Pulsars," *The European Physical Journal: Historical Perspectives on Contemporary Physics* 38, no. 4 (2013): 535–47.
3. I came across this work at the tail end of writing this book, after this section was complete, but felt it appropriate to cite that others have made similar arguments in astrobiology regarding the Sagan Standard: Sean McMahon, "Do Extraordinary Claims Require Extraordinary Evidence?" *Social and Conceptual Issues in Astrobiology* (2020): 117.

4. Abhishek Sharma et al., "Assembly Theory Explains and Quantifies Selection and Evolution," *Nature* 622 (2023): 1–8.
5. Everett Shock et al., "Earth as Organic Chemist," in *Deep Carbon: Past to Present*, ed. B. N. Orcutt, I. Daniel, and R. Dasgupta (Cambridge, UK: Cambridge University Press, 2019), 415–46.
6. Steven A. Benner, "Rethinking Nucleic Acids from Their Origins to Their Applications," *Philosophical Transactions of the Royal Society B* 378, no. 1871 (2023): 20220027.
7. Caleb Scharf and Leroy Cronin, "Quantifying the Origins of Life on a Planetary Scale," *Proceedings of the National Academy of Sciences* 113, no. 29 (2016): 8127–32.
8. Alan Ianeselli et al., "Physical Non-Equilibria for Prebiotic Nucleic Acid Chemistry," *Nature Reviews Physics* 5, no. 3 (2023): 185–95.
9. Robert M. Hazen and Dimitri A. Sverjensky, "Mineral Surfaces, Geochemical Complexities, and the Origins of Life," *Cold Spring Harbor Perspectives in Biology* 2, no. 5 (2010): a002162.
10. Andrew J. Surman et al., "Environmental Control Programs the Emergence of Distinct Functional Ensembles from Unconstrained Chemical Reactions," *Proceedings of the National Academy of Sciences* 116, no. 12 (2019): 5387–92.
11. Nigel Goldenfeld, *Lectures on Phase Transitions and the Renormalization Group* (New York: CRC Press, 2018).
12. Geoffrey B. West and James H. Brown, "Life's Universal Scaling Laws," *Physics Today* 57, no. 9 (2004): 36–42.
13. Dylan C. Gagler et al., "Scaling Laws in Enzyme Function Reveal a New Kind of Biochemical Universality," *Proceedings of the National Academy of Sciences* 119, no. 9 (2022): e2106655119.
14. Searra Foote, Pritvik Sinhadc, Cole Mathis, and Sara Imari Walker, "False Positives and the Challenge of Testing the Alien Hypothesis," *Astrobiology* 23, no. 11 (2023): 1189–1201.
15. Sara Imari Walker, "Origins of Life: A Problem for Physics, a Key Issues Review," *Reports on Progress in Physics* 80, no. 9 (2017): 092601.
16. Philip Morrison and Giuseppe Cocconi, "Searching for Interstellar Communications," *Nature* 184, no. 4690 (1959): 844–46.
17. Frank D. Drake, "Project Ozma," *Physics Today* 14, no. 4 (1961): 40–46.
18. Brandon Carter, *Anthropic Principle in Cosmology* (Cambridge, UK: Cambridge University Press, 2006).

19. Brandon Carter, "The Anthropic Principle and Its Implications for Biological Evolution," *Philosophical Transactions of the Royal Society of London. Series A, Mathematical and Physical Sciences* 310, no. 1512 (1983): 347–63.
20. Jane S. Greaves et al., "Phosphine Gas in the Cloud Decks of Venus," *Nature Astronomy* 5, no. 7 (2021): 655–64; Antonin Affholder et al., "Bayesian Analysis of Enceladus's Plume Data to Assess Methanogenesis," *Nature Astronomy* 5, no. 8 (2021): 805–14.
21. G. L. Villanueva, Martin Cordiner, P. G. J. Irwin, Imke de Pater, Bryan Butler, Mark Gurwell, S. N. Milam, et al., "No Evidence of Phosphine in the Atmosphere of Venus from Independent Analyses," *Nature Astronomy* 5, no. 7 (2021): 631–35.
22. William Bains et al., "Phosphine on Venus Cannot Be Explained by Conventional Processes," *Astrobiology* 21, no. 10 (2021): 1277–1304.
23. Giada N. Arney et al., "Pale Orange Dots: The Impact of Organic Haze on the Habitability and Detectability of Earthlike Exoplanets," *Astrophysical Journal* 836, no. 1 (2017): 49.
24. Harrison B. Smith and Cole Mathis, "Life Detection in a Universe of False Positives: Can the Fatal Flaws of Exoplanet Biosignatures Be Overcome Absent a Theory of Life?" *BioEssays* (2023): 2300050.

5: ORIGINS

1. Stanley L. Miller, "A Production of Amino Acids Under Possible Primitive Earth Conditions," *Science* 117, no. 3046 (1953): 528–29.
2. H. James Cleaves II, Grethe Hystad, Anirudh Prabhu, Michael L. Wong, George D. Cody, Sophia Economon, and Robert M. Hazen, "A Robust, Agnostic Molecular Biosignature Based on Machine Learning," *Proceedings of the National Academy of Sciences* 120, no. 41 (September 25, 2023), https://doi.org/10.1073/pnas.2307149120.
3. Alan M. Turing, "Computing Machinery and Intelligence," *Mind* 59, no. 236 (1950): 433.
4. Alan Mathison Turing, "On Computable Numbers, with an Application to the Entscheidungsproblem," *Journal of Mathematics* (1936): 345–63.
5. J. von Neumann, *Theory of Self-Reproducing Automata*, ed. Arthur W. Burks (Urbana: University of Illinois Press, 1966).
6. S. Mehr et al., "A Universal System for Digitization and Automatic Execution of the Chemical Synthesis Literature," *Science* 370, no. 6512 (2020): 101–108.

7. Silke Asche et al., "A Robotic Prebiotic Chemist Probes Long Term Reactions of Complexifying Mixtures," *Nature Communications* 12, no. 1 (2021): 3547.
8. Laurie J. Points et al., "Artificial Intelligence Exploration of Unstable Protocells Leads to Predictable Properties and Discovery of Collective Behavior," *Proceedings of the National Academy of Sciences* 115, no. 5 (2018): 885–90.
9. Juan Manuel Parrilla Gutierrez et al., "Evolution of Oil Droplets in a Chemorobotic Platform," *Nature Communications* 5, no. 1 (2014): 5571.

6: PLANETARY FUTURES

1. Some of the material in this chapter previously appeared in: Sara Walker, "AI Is Life," *Noēma*, April 27, 2023, www.noemamag.com/ai-is-life.
2. Robin Hanson, "The Great Filter—Are We Almost Past It?" (1998), https://mason.gmu.edu/~rhanson/greatfilter.html.
3. David J. Chalmers, *Reality+: Virtual Worlds and the Problems of Philosophy* (London: Allen Lane, 2022).
4. James E. Lovelock and Lynn Margulis, "Atmospheric Homeostasis by and for the Biosphere: The Gaia Hypothesis," *Tellus* 26, nos. 1–2 (1974): 2–10.
5. John Maynard Smith and Eörs Szathmáry, *The Major Transitions in Evolution* (Oxford, UK: Oxford University Press, 1997).
6. Adam Frank, David Grinspoon, and Sara Walker, "Intelligence as a Planetary Scale Process," *International Journal of Astrobiology* 21, no. 2 (2022): 47–61.

INDEX